Risk Management

PRACTICAL FARMING

Rural Property Planning
Risk Management

Mike Krause

INKATA PRESS

INKATA PRESS
A DIVISION OF BUTTERWORTH-HEINEMANN

AUSTRALIA
BUTTERWORTH-HEINEMANN
North Tower 1–5 Railway Street
Chatswood NSW 2067

BUTTERWORTH-HEINEMANN
18 Salmon Street
Port Melbourne 3207

UNITED KINGDOM
BUTTERWORTH-HEINEMANN LTD
Oxford

USA
BUTTERWORTH-HEINEMANN
Stoneham

National Library of Australia Cataloguing-in-Publication entry

Krause, M. A. (Michael Andrew).
Rural property planning: risk management.

Includes index.
ISBN 0 7506 8917 X.

1. Farm management - Australia. I. Title. (Series: Practical farming).

630.68

Typeset by Ian MacArthur, Hornsby Heights, NSW.

Printed in Australia by Ligare Pty Ltd, Riverwood, NSW

Contents

1 What is risk?

Risk implies hazard, uncertainty, fear, gambling and chance. None of these ideas are particularly appealing to people trying to run a business such as a farm or grazing property. Risk is therefore something we try to control and suppress.

People who can manage risk are generally better able to manage their farms by keeping the stresses associated with risk where they belong—a prod to perform better—not something that causes health and business problems.

It is important to remember that risk is a fact of business life and generations of Australian farmers have lived with risk since agriculture first began here.

Risk is changing in the Australian rural sector. Now financial and stress risks are more important than they were years ago, while physical injury risks are generally becoming less important. As a result, more financial management effort may now be needed to measure and manage risk. Farmers are therefore developing new tools to help better manage these new risks. Computers for farming are one example.

People use the term risk quite loosely in agricultural industries. For example, some people believe risk relates mainly to unexpected fluctuations in commodity prices, while others see it as the consequence of varying weather on crop production or on the health and well-being of farm families. In reality, risks that impact on farms may arise from many factors.

Farmers manage risk in different ways and with varying degrees of success. Everyone has heard of instances where families have had to leave the land because debts got too high, or drought dried up farm incomes, or prolonged downturns in commodity prices meant farms were no longer viable. These management problems often occurred because farmers first, did not recognise the effects of risk and second, failed to manage their farms accordingly. However, if risk is properly managed there is a good likelihood farms will remain successful.

Risk or uncertainty

Risk and uncertainty are defined as two different things:

- ❑ risk is the estimated measure, or probability, of something happening
- ❑ uncertainty covers all things that could happen which the decision maker would never have expected.

In recent times these two terms have been combined in the one definition of risk. For example we can estimate this season's cereal yield will probably be three tonnes per hectare. This probability will give us a measure of risk. In fact the harvest may get unseasonal rainfall which means the crop quality is downgraded. This event would be completely unexpected and would also need to be assessed within the framework of risk. Risk deals with predictable variation and completely unexpected events.

Farming risk in perspective

There are many risks that confront farm businesses.

Market risks include:

- ❑ commodity price fluctuations—such as changing grain prices
- ❑ exchange rate variations—which can have a big impact on export commodity prices
- ❑ cost variations due to inflation—affecting input costs such as fuel and chemicals
- ❑ the long term trend of declining terms of trade—which can erode farm profits.

Financial risks include:

- ❑ interest rate shifts—these have an obvious effect on debt repayment
- ❑ declining equity—which makes it harder for businesses to cope with adverse seasons
- ❑ land value fluctuations.

Production risks include:

- ❑ continuity and supply of labour. Labour supply can be crucial during critical operations such as harvesting, spraying and planting.
- ❑ managing fertiliser, seeding and applying chemicals. This can have a big influence on final production and hence yields.
- ❑ land management and land conservation can have a critical impact on the land's long term productivity
- ❑ managing animals

- ❑ selecting cultivars—can influence crop productivity
- ❑ seasonal rainfall has an effect on yield and has obvious financial impacts on profits
- ❑ diseases—can have an impact on crop yields and livestock productivity
- ❑ mechanical breakdowns—can severely affect farm operations and productivity
- ❑ theft
- ❑ disasters such as hail, fire and flood all have a direct impact on cash flow and profitability by damaging assets and productivity.

Personal risks include:

- ❑ manager and staff health; physical health, mental health and stress
- ❑ family support and harmony may influence both labour supply and decision making capacity
- ❑ community support—can impact on the property's productivity, especially in times of disaster or personal crisis
- ❑ business succession. This is often a difficult area for farming families and the property's long term financial survival is very dependent on the way successions are handled, for example marriage breakdown and the fate of older generations in retirement.

Management risks:

- ❑ problems caused by poor or inadequate information can directly affect farm productivity
- ❑ keeping abreast of changes in technology can often have a big impact on farm profits and viability. For example, failing to switch to sensible supplementary urea feeding programs can greatly reduce the carrying capacity of some cattle grazing properties in northern Australia.

These are just some of the risks that can effect farming. Obviously the influence of these risks are not all equal, as some will have a larger potential effect on profits and assets than others. There will also be variation between seasons and over the long term, as the economy, people's knowledge and society changes.

Sensitivity analysis

To illustrate which risks are more important than others, a farm consultant carried out a sensitivity analysis on a farm on South Australia's Yorke Peninsula (Table 1.1).

In this example the consultant worked on 5 per cent changes.

It is important to remember that these analyses vary from case to case, so the 5 per cent level used here is for illustration purposes only. The analysis shows those factors that have the biggest financial influence on profits.

Table 1.1 *Sensitivity analysis: Effect on net profit of a five per cent change in value (before tax)*

		Original Value	*New Value*	*Change in Value*	*Net Profit $*	*Ranking*
exchange rate	$US/$A	0.74	0.7	0.04	5311	1
feed barley yield	t/ha	2.58	2.71	0.13	2430	2
feed barley prices	$/ha	105	110.25	5.25	2411	3
wheat price	$/ha	130	136.5	6.5	1620	4
wheat yield	t/ha	2.8	2.94	0.14	1620	5
interest rates	%	11.75	11.16	0.95	1479	6
machinery ownership cost	$	25022	23771	1251	1251	7
living expenses	$	20000	19000	1000	1000	8
wool production	kg	5293	5557.65	265	925	9
wool price	c/kg	3.5	3.68	0.18	884	10
pea price	$/ha	200	210	10	640	11
pea yield	t/ha	2	2.1	0.1	640	12
lambing %	%	64.2	67.4	3.2	635	13
fuel costs	$	9659	9176	483	483	14
fat lamb prices	$/ha	22.5	23·63	1.13	470	15
livestock costs	$	8426	8005	421	421	16
fertiliser prices	$	7937	7540	397	397	17
repairs and maintenance	$	5384	5115	269	269	18
chemical costs	$	3488	3314	174	174	19
insurance	$	816	775	41	41	20

Note: this farm business has 68 per cent equity. If this business had a 90 per cent equity interest rates would be ranked 13 in the table.

The seventh ranked item, 'machinery ownership cost' would normally include factors such as the opportunity cost of interest. This opportunity cost is the interest foregone by having money tied up in machinery instead of holding it in a bank account or similar investment. To keep the table simple, we have not included such items.

This sensitivity analysis highlights the effect on the property's net profit of a 5 per cent change in each of the individual factors listed in the table above. Each factor has been changed independently of the other factors by 5 per cent. From this, the consultant worked out the change in net profit. Then the consultant ranked the factors in descending order of importance.

For example, a 5 per cent shift in the $A/$US exchange rate from $US 0.74 to $US 0.70 caused profits to increase by $5311. This mainly happens because of the effect exchange rate movements have on grain prices. As this was the largest effect on profits caused by any factor, the exchange rate got the highest ranking.

While management can directly affect some factors, the most important are largely beyond the control of managers. As a result, managers need to cope with these factors by using several risk management strategies, not one.

Seasonal factors obviously have a large influence on this example property's profitability. That is, shifts in barley and wheat yields ranked quite high on the sensitivity rankings.

Those areas that are influenced by outside economic factors include exchange rates, grain prices, interest rates and wool prices. These are among the most sensitive factors and so present the biggest risks to the property's profits. In later chapters we will see management tools and strategies that allow business managers to cut the risks of commodity price movements.

While sensitivity analysis is not a perfect way of analysing risk, it does indicate the biggest potential risks to a farm's finances. It does highlight the risk areas on which a manager needs to concentrate.

Risk and the current economic situation

The economic recession of the early 1990s posed its own risks on farm businesses. Some of these risks are listed below.

Economic deregulation

Recent increases in risk have been largely due to deregulating parts of the domestic economy. Examples include:

- ❑ deregulating the financial industry
- ❑ phasing out the guaranteed minimum price for wheat
- ❑ partially privatising the Commonwealth Bank
- ❑ amalgamating Qantas and Australian Airlines
- ❑ floating the Australian dollar ($A).

These changes have either caused direct or indirect risk increases in agriculture during recent years.

An example of added deregulation due to market failure is in the wool industry. Wool industry deregulation resulted in the suspension of the Australian Wool Corporation's wool floor price. This was an example where previous regulation was mismanaged and went wrong, leaving deregulation as the only viable option for the wool industry.

The Federal Government's deregulation policy of the last ten years has changed the way economic activity is being undertaken in Australia. Rural industries have been one of the first industries involved in deregulation. Some analysts believe deregulation started in agriculture because it was easier to manage than in more labour-intensive industries, like the car industry.

The major factors that have increased farming risk in recent times include:

- ❑ deregulating the wheat industry
- ❑ the removal of the floor price for wool
- ❑ relaxing rules governing the finance market and exchange rates
- ❑ the increased trade war between the EC and the US which has made international markets more unpredictable and increased the difficulty of profitably marketing goods.

Economic recession

While most farm incomes are dependent on international markets, there are some increased risks being felt by farmers due to Australia's domestic recession. These risks are especially important in cases where farmers are thinking of leaving the land. In these cases the recession has cut alternative ways of earning a living, due to high unemployment rates and the high numbers of small business bankruptcies.

However, the current recession has caused two advantages:

- ❑ inflation is lower
- ❑ interest rates are down.

While fluctuating interest and inflation rates add to business risk, sometimes these variables work in the favour of business. 1994 is one of those occasions.

Risk from seasonal variation

One of the major causes of uncertainty is seasonal variation. This has two big impacts on risk:

- ❑ overseas seasonal variation, which influences overseas production
- ❑ domestic seasonal variation, which influences domestic production.

Overseas seasonal variation

World supply and demand for commodities can be influenced by a number of factors. The biggest of these is weather. If the world has low stocks of a rural commodity and one of the main world suppliers has a major drought, short supplies may trigger an increase in the world price.

A clear example of this happened in 1987/88 when the US had a poor wheat season. The poor US crop caused world wheat shortages which made the world wheat price rise dramatically.

Another example is the effect of drought on Brazilian citrus juice production. As Brazil is the world's largest citrus juice producer, when it has a drought, world supply falls and the price of citrus juice goes up. Weather variations in Brazil can therefore cause unstable prices for citrus producers around the world.

As global seasonal variations occur, there always will be some uncertainty in supply. This uncertainty will lead to price variations on international markets.

Domestic seasonal variation

Like the global affects of seasonal changes, local seasonal effects also influence the Australian domestic market.

An excellent example is the vegetable industry. Vegetable growers generally get high prices when some other competing region within Australia has excessive hail, frost or floods which disrupt their supply. Similarly when drought occurs, farmers need extra fodder. This means the price of hay and grain generally increases.

As long as we have unpredictable seasonal variations, there will always be uncertainty in rural commodity markets.

Other catalysts of uncertainty

Further influences of uncertainty that will affect the medium to long term markets for Australian rural exports are:

- ❑ the trade war between the EC and the US
- ❑ the economic developments in eastern Europe
- ❑ the continuing economic developments in south-east Asia.

Trade wars

Current uncertainty in world agricultural trade has been worsened by the subsidy trade war between the EC and the US, particularly in the world's grain trade. Recent General Agreement on Tariffs and Trade (GATT) talks have improved the trade situation between the US and the EC, and these agreements have now been endorsed by Congress in the USA.

Trade wars are selective and hit at different world markets. While they remain, world commodity prices will stay depressed, particularly for grains. This creates continuing price uncertainty for Australian farmers.

Economic developments in eastern Europe

The breakdown of communism in the former Soviet Union and eastern Europe has resulted in major economic disruption in those economies.

While there is still political unrest and war in some of the emerging eastern European countries, these nations have the potential to directly compete with Australian exports.

Once their unrest is overcome, these countries will be initial importers of rural commodities, but once they have developed competitive infrastructures and have increased their agricultural investment, they may compete with Australia in international markets. They may take ten to fifteen years to compete with us, but when they do they will bring increased competition and uncertainty to Australian farms.

Economic developments in south-east Asia

We have seen strong economic development in Japan, Singapore, Taiwan and Korea over recent times. This increased economic growth and standard of living is now spreading to other Asian countries such as Indonesia, Thailand, Vietnam and China.

There is an advantage for Australian rural exports in this development. As Asia's standard of living increases, the Asian diet will become more westernised. Hence there may be an increased demand for cereal grains, meat, fruit, vegetables, and dairy products. These countries may try to produce these commodities themselves, but they will soon meet resource limitations. This will force them to import to satisfy the increased demand for these products. As this occurs, Australia will be in an excellent position to expand its farm exports.

Understanding manageable risks

This book covers most farm risks. It also outlines strategies and management techniques that can be used to better manage risk. Some risks lie within managerial control, such as fire, while others are beyond managerial influence, like currency exchange rates. The art of sound risk management is to manage those risks that can be managed and to minimise the effects of those risks that cannot be managed.

2 The farm business system

One of the aims of farm business management is to assess the profit levels of various farm activities and strategies. This is usually done by using a series of budgets. The assessment of farm risk is basically an extension of farm management measures designed to measure profit and changes in the value of farm assets.

The farm business risk framework

As outlined above, there is a broad range of risks that can influence the farm business. The best way to describe, then analyse, these risks is to develop a framework for those risks that may affect a farm business. The flow chart (Figure 2.1) illustrates this framework.

The risk framework of a farm business as shown in the chart, is prepared in three main sections:

- external risk factors
- management
- the farm.

Figure 2.1 shows the external risks occurring independently of the farm business. These risks impact on the farm business both singularly and collectively.

The management component is essential when it comes to assessing how much risk is allowed to affect the farm business. Obviously some risks are unavoidable, while others can be more easily managed. However, as will become evident throughout this book, it is what management does with the risks that will determine how financially successful the property becomes.

The risk framework chart shows the farm is made up of two major components:

- production resources
- financial results.

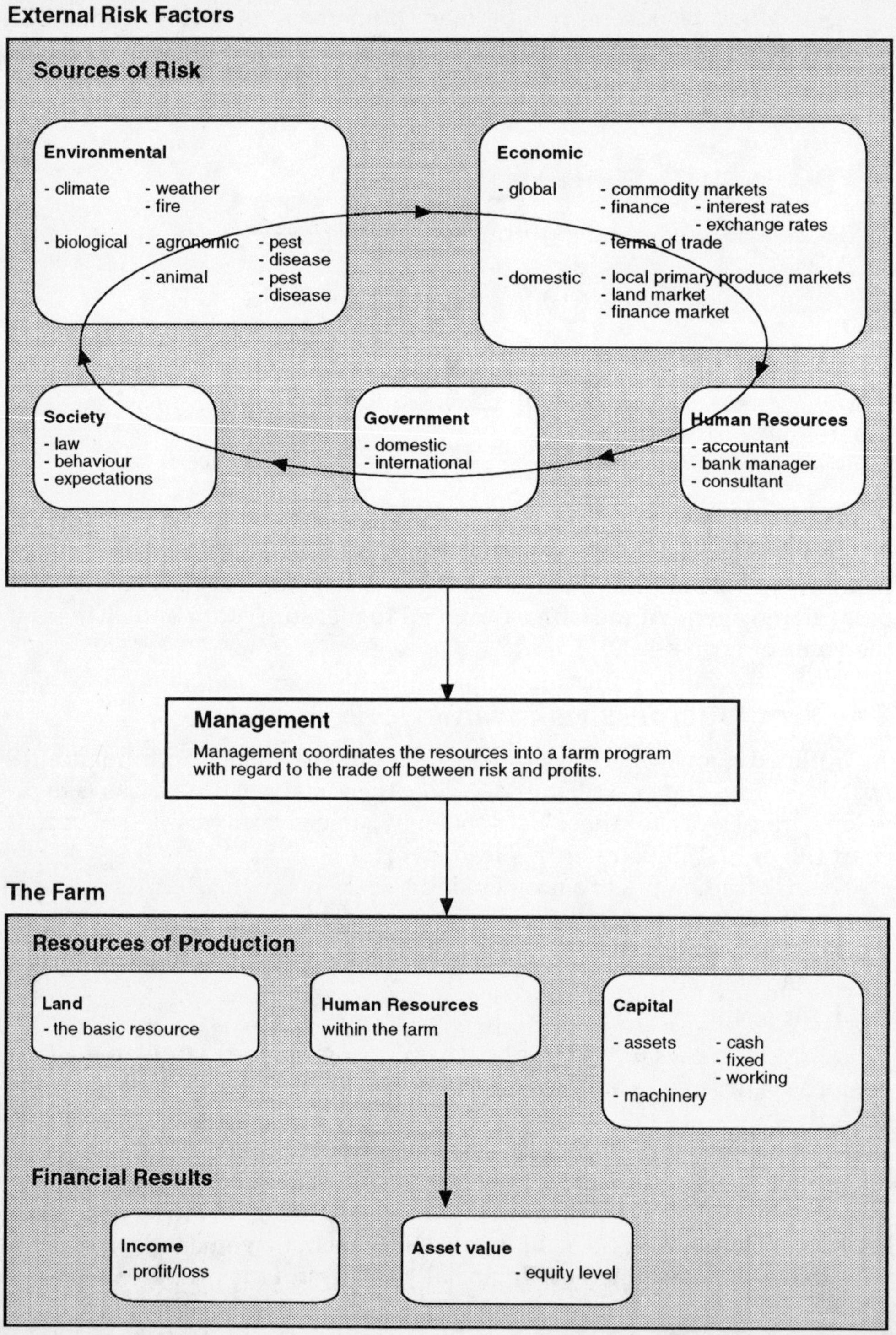

Figure 2.1 *Farm business risk framework*

Production resources include land, human resources and capital. The management of risk in these three areas will determine the level of income and asset value generation that will occur over time.

Production resources

Economists have described the farm (or rural business) as being made of:

- ❑ land
- ❑ human resources
- ❑ capital.

These are seen to be the basic ingredients that create a rural business. It is management that works with these three components to create the financial return for the business.

Land

As a component of farming, the impact of land on business viability is easy to understand. For broad-acre, grazing and horticultural properties, land is vital. For piggeries and poultry businesses it is less vital, but is still an important component.

When considering the impact of land on risk the following considerations should be taken into account:

1. Land is a finite resource and is usually the most restrictive resource on cropping and grazing properties. That is why most gross margins are expressed in terms of dollars per hectare. Being the most finite resource means the profits need to be maximised to the amount of land available.
2. Land is usually the most expensive resource in a farm business. While its value does fluctuate, the fluctuations tend to be smaller and lag behind commodity prices and seasonal trends.
3. Since land is the most valuable business resource, it is usually the one resource that banks will take as a mortgage against any loans for the business. Hence land is a vital security component of the farm business.

In recent years land sustainability and land care issues have become topical. These are important issues and address the problems of over-exploitation of land and the continual degrading of its quality by factors such as erosion, salt or falling nutrient levels. Land care highlights some risks associated with land use. For example short term exploitation of the land may result in long term loss.

HINT	
H I N T	Some alternatives to owning land are: ❑ share farming ❑ leasing land.

Human resources

Human resources are people resources or inputs. These people inputs are usually made up of either physical labour or managerial capacity. Both are necessary to run a successful farm.

Physical labour

Skilled labour and how much you can get when you need it are all factors that affect the risk of farming. The greatest wine grape crop, the best hay cut, the best cattle production may all be for nothing if labour is not available to pick the grapes, bale and cart the hay, or muster the cattle.

Labour can come from the following areas:

- owners' labour
- family members' labour
- hired labour.

Most farming operations are managed by family units and therefore the family supplies most of the labour.

The requirement for labour varies between farming enterprises, with generally more labour needed in more intensive agricultural activities—such as piggeries, poultry farms and glasshouse vegetable farms.

Managerial capacity

The skills and experience of farm managers are other important risk areas. In equal circumstances, skilled and experienced managers will generally manage risk better than unskilled or inexperienced farm managers.

It is important to distinguish between labour and management.

Although they are both resources people bring to a business, they are often quite separate skills. Although they may often be carried out by the same people (a farmer/manager), they need to be treated as separate items when assessing risk—just as drought and changing commodity prices need to be assessed separately.

The flow of people from the rural areas to town and city areas continually concerns rural communities. The continual economic pressure on farming is only making this situation worse. It is a problem facing an increasing number of rural communities, because skilled rural labour is becoming less and less available for rural businesses.

Capital

Like land and people, capital (money) is subject to business risks.

Capital consists of several things:

- land
- improvements that have been added to the land, such as houses, watering systems, fences and sheds
- machinery
- cash reserves.

Capital can be broken up into a number of areas:

- liquid reserves—such as cash in the bank or shares and similar assets that can quickly be turned into cash
- fixed assets
- working assets—such as machinery, livestock, fodder, produce.

Cash assets

Cash assets are those funds that can be readily accessed in times of need. These are the funds that can be drawn on during hard times. Examples where farms may need to draw on cash reserves include poor wool returns, droughts, and family sickness.

Obviously a sizeable cash reserve greatly assists any risk management strategy in times of financial stress.

The risk management strategy of keeping quite sizeable cash reserves has been relatively common over the history of Australian agriculture. The fact that a third of farm businesses currently are debt free, indicates this strategy is still widely used.

Fixed assets

Fixed assets, otherwise known as improvements, are items that have been added to the property. These include items such as fences, buildings, the house, watering systems, waste disposal systems, stock yards and irrigation layouts.

Fixed assets add value to the property and enable the respective enterprises to be conducted more efficiently.

These items are known as fixed assets, as they cannot be easily detached from the property and sold separately.

Working capital

Working capital is the cash generally required to operate day to day activities on the farm. It is used to buy inputs such as seed, fertiliser, fuel, chemicals and veterinary needs. Most farmers get supplies of working capital by using the overdraft facilities of a bank.

There is a subtle difference between working capital and cash. Cash assets are usually used to accumulate wealth and are preserved for emergencies. Working capital may be a combination of borrowed overdraft and the owner's own money.

Machinery

Capital is required to buy machinery. Machinery is essential for most properties as it allows farmers to make best use of available labour and also enables many farm operations to occur at the most suitable times.

Machinery also needs to be maintained and replaced when necessary.

Therefore farmers need to keep some sort of replacement fund in place so they can buy replacement machinery when it becomes necessary.

One big problem facing Australian farmers is the increasing age of on farm machinery and the lack of money to replace it. The tight financial situation of recent years, coupled with high machinery prices, has meant farmers have not been replacing machinery. These circumstances have introduced several risks to Australian farms:

- ❑ the cut in machinery replacement means farm operations may become less reliable. The timeliness of operations may also be affected and business productivity may fall.
- ❑ ageing machinery increases farmer anxiety. Many farmers feel they have little chance of replacing machinery they see as being essential to maintain their businesses in the long term. In recent times many farms have only been able to maintain a positive cash flow by not replacing machinery.
- ❑ a collective run down of machinery means an industry cannot work as efficiently.

One possible solution to letting machinery run down because of lack of money is to look at using other means for using machinery. These other means include using contractors, setting up an ownership syndicate or hiring machinery. These options may help reduce the overhead costs that come with owning machinery and therefore free some money.

HINT	Some alternatives to owning machinery are: ❑ using contractors to do some operations ❑ forming a machinery syndicate with other farmers ❑ hiring machinery.

Financial results

The second major component of the farm business is its financial record. Financial results are a measure of financial health. If, over time financial results are deteriorating, these will increase the risk of other factors hitting the farm hard. Conversely, if the financial results improve over time, the property will be in a better situation to withstand any financial disasters in the future. Wealth will not affect business uncertainty, but there is far less risk in farming with healthy financial results than with poor ones.

The two financial reports that can indicate the possible effects of risk are:

- ❑ the annual whole farm budget—which measures cash flow
- ❑ the asset and liability budget—which shows if net assets are growing or shrinking.

The profit and loss budget

The annual influence of risks is typically shown through price variations, yields and changes in costs. All three factors have a direct influence on profits, so the calculation of the annual whole farm budget is an excellent annual indicator of the impact of risk on the property.

An example of a whole farm budget is shown in Table 2.1. This whole farm budget shows the return to capital (profits) of $17 679. That is the season's yields, prices and costs produced a profit of $17 679.

It is possible to generate a risk profile of this property's potential profit results. As outlined in Table 2.1, most risk impact on a property is through variations in prices, yields and costs.

Most farmers expect to get a price somewhere in a likely price range. Likewise, farmers will expect yields to be somewhere within a given range and expect similar ranges for costs.

At the beginning of a season, it is very difficult to be certain how the coming year will turn out financially. However, with a special computer programming technique, it is possible to estimate a likely financial result. This technique calculates the likely results of 1000 seasons. Each time the computer calculates a seasonal result, it picks a new set of yields, prices and costs from the expected range then calculates these.

After a 1000 season calculation, an excellent impression can be gained on the likely financial result of the property and its risk profile. The risk profile for a farm is shown in Figure 2.2. The results in this graph indicate the likely spread of income. On a 1000 season average, the most probable profit was $6100 and not the $17 679 as indicated in the whole farm budget in Table 2.1. The difference was caused because the program made allowances for average variations in prices, yields and costs. The interaction of these prices, yields and major costs caused the drop in expected profits.

This computer technique averages out variable factors over many farms and/or seasons and so produces a more reliable average. It will not tell

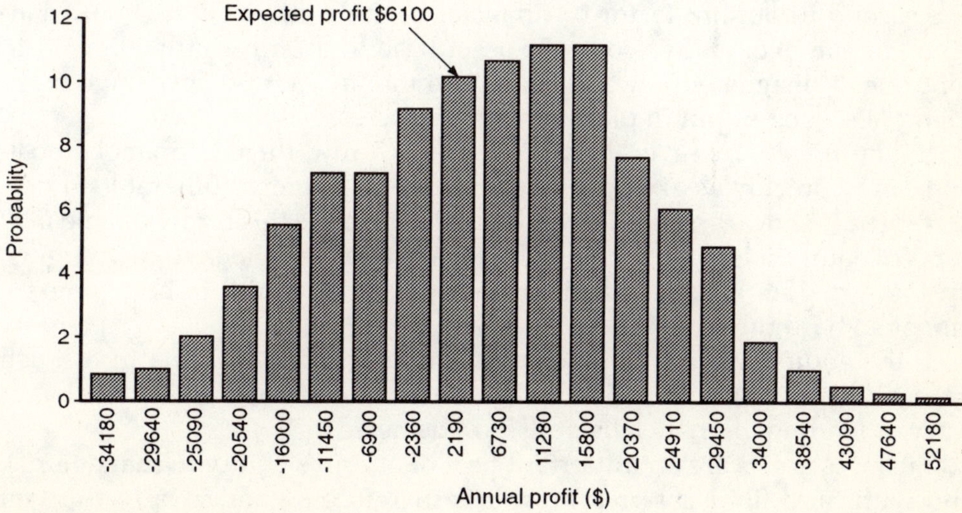

Figure 2.2 *Expected profit variation*

Table 2.1 *Whole farm budget*

Total Farm Income	wheat	$14 040
	barley	57 877
	peas	26 306
	vetch	17 689
	wool	9 065
	sheep	17 100
	Total	142 077
Total Costs	cropping & past	26 310
	livestock	9 560
	overheads	32 500
	interest	9 048
	wages	2 200
	Total	79 618
Farm Operating Surplus		62 459
Depreciation (machinery replacement)		24 780
Operator/management fee		20 000
Return to capital (profit)		17 679
		2.3%
Farm business equity ratio		89.9%
Walk in walk out value		= 761 445
Farm debt		= 77 000

which year is going to be good and which will be bad. Farmers with many years experience in an industry or in a district might regard these results as rather obvious, but the technique may have a place in assessing risk on your farm.

Asset value

The second important factor of financial health is asset value. Poor management can eventually erode the asset base of a farm if it is allowed to continue. Management always should be monitored and corrective management strategies put in place whenever necessary.

If a farm is hit by several poor seasons in a row, then its financial position can be quickly weakened and the farm will be more vulnerable to risk.

Table 2.2 shows assets and liabilities of a farm. As this farm has a near 90 per cent equity, it is in a good position to take financial shocks like droughts, fires and market collapses. This property has the equity to borrow extra money in hard times.

If the equity was only 60 per cent, the property would be in a much tighter financial situation if it had to borrow more, as banks are always wary of lending to financially stressed businesses.

Business assets are an important part of a farm's ability to manage risk and withstand financial pressures. It is therefore a good idea to think of strategies that might be needed to maintain your business assets.

Table 2.2 *Assets and liabilities*

Assets	***$***	***Liabilities***	***$***
Land & improvements		Bank overdraft	22 000
		Term Loan	55 000
245ha @ $1235/ha	599 245		
319ha @ $930/ha			
Machinery	130 000		
Livestock		Other debts	
sheep	8 000		
Stores			
Equity in Barley pools	4 200		
Investments	20 000		
Total Assets	761 445	Total liabilities	77 000
Equity	684 445	Equity %	89%

3 Weather-related risks

The vagaries of nature play a big part in primary production, so nature and its variables create some of the biggest risk factors in farming.

Management has little or no direct influence over the effects of nature while nature can have extreme effects on profits. Good weather often means good income, as long as prices are good, but bad weather can hit profits hard.

Nature is a bit too large a subject to understand easily. That is why many managers and consultants break it down into its various parts for management purposes. These parts include:

- climate—which is made up of drought and less extreme weather factors
- biological factors.

By dividing the environment into parts it is easier to analyse the risks each pose to farming.

Weather risks are major risks to farming. Damage caused by hail and fire are two examples of how weather can physically and financially hit farm businesses.

Deciding how much effort should go into managing weather related risks is often difficult and will vary greatly from farm to farm. One approach would be: "If I can't afford weather related losses then I had better manage that risk".

Rain

Rain is one of the most valuable resources for most farms and grazing properties. Adequate rain provides good seasons, which lead to bigger profits. On the other hand, too much (or too little) rain at certain times of the year can greatly affect productivity. Events such as drought can be very devastating to livestock, crops and farm finances. Too much rain also can cause major problems in farming operations, especially at harvest time where too much moisture often results in spoiled crops. Unseasonal rain can greatly reduce returns to farmers.

Two recent examples of unseasonal rain are:

- in far north Queensland in 1990/91 torrential rain and strong winds from a tropical cyclone damaged banana and sugar crops from Mossman to Innisfail
- in 1992/93 heavy unseasonal rain in South Australia caused grain to sprout in the head before many crops could be harvested. The rain turned potential record harvests into only poor to average ones.

Excessive moisture at harvest is also a problem in cotton crops, where the cotton can discolour and drop in quality. Harvesting also can be delayed as machinery cannot operate in the wet and boggy conditions that occur on the black soils of the cotton country around Moree and Narrabri. Likewise, wine and table grapes and many other types of fruit can suffer in quality if unseasonal rain affects fruit—either by physical damage or causing diseases and fungi to grow.

Hail also can be very damaging, causing total destruction of fruit and cereal crops in southern Australia. Some farming areas are more prone to hail damage than others.

Practical aspects of rainfall risk

There are several practical aspects to managing the risks associated with rainfall. Some of these are:

- too much rain (floods)
- too little rain (drought)
- timing of the rain (unseasonal falls)
- the form it takes (hail).

Drought is covered in detail in Chapter 4, but it is worth mentioning here that drought insurance is an additional strategy that has been used by large corporate agribusinesses. This strategy may become more available to other farmers in the future.

Excessive rainfall

Too much rain is a risk that sometimes reduces the value of crops and livestock. Taking out insurance is a common way to cover these losses. The list below shows some insurable examples:

- cotton—can suffer from a drop in quality after rain
- sugar cane—also can suffer from a drop in quality
- tomato growers—have trouble putting harvest machinery onto land after heavy rainfall

- wheat crops—too much rain at harvest can downgrade quality
- barley, especially malting barley—suffers quality problems with excessive rainfall
- cherries—suffer from splitting if high levels of rainfall occur at harvest
- grapes—excessive rain can cause quality and yield losses
- flowers—growers can insure against damage to blooms caused by excess rain.

Other high rainfall risk management strategies might include land care strategies designed to reduce soil degradation from rain and water. Strategies here could include contour banks and laser-levelling paddocks.

Hail

Hail is another hazard where farmers can cut risks, including insuring against losses.

Insurance is a widely available and commonly used strategy to manage the risks hail poses to farming operations. Insurance cover is available for:

- all broad-acre crops including cereals and grain legumes
- citrus
- vegetables
- grapes
- flowers
- nuts.

Other strategies to help cut hail risks are:

- hail netting for horticultural crops
- hail cannons—while not yet proven, some horticulturists use these devices, which are claimed to prevent hail forming during stormy conditions
- susceptible crops spread over a large area to reduce the likelihood of one hail storm affecting all crops
- to select less susceptible crops, such as growing almonds rather than apricots.

Temperature

The main effects of temperature in Australian agriculture are poor growth of livestock and crops because of cold and heat stress on plants and animals. Frost and heat stress can be quite devastating to some enterprises. In crops, cold conditions, particularly frost, can prevent or reduce fruit-set in stone fruit and grapes as well as seed-set in cereals. In some crops such as cereals, frost damage is not evident until quite close to harvest when grains do not fill.

Excessive rainfall can hold up vital operations

Practical aspects of cold risk

Low temperatures require a range of measures to minimise the risk they pose to farming operations. Some risks low temperatures pose to farming include endangering the lives and/or growth rates of:

- small animals, such as chicks, bees, piglets, calves and lambs
- newly-shorn sheep
- poultry.

Farmers can minimise risks by:

- providing heating for at-risk animals
- insulating sheds
- providing shelter and windbreaks for free-range stock.

Plants too can benefit by being protected from cold conditions. An example is tomato growing, where greenhouses protect plants from the cold.

This allows tomato seedlings to grow into warmed greenhouses months before such plants can go into outside beds. This practice ensures early harvests with subsequent higher prices. Growers can minimise most weather risks, because the climate inside many modern greenhouses can easily be controlled.

Examples are:

- ❑ insulating greenhouses by lining the main structure with a double glazing layer of polythene or similar film
- ❑ providing either solar or fuel heating at periods of peak cold—gas heaters are often used in the early hours of the morning, the coldest time of day.

Practical aspects of frost risk

Frost is one of the more common risks associated with cold temperatures in much of southern Australia. There are a number of strategies farmers can use to help manage the financial risk of frost. One of the most common is to take out frost insurance. Frost insurance is available for:

- ❑ tree crops, including pome fruit, stone fruit, nuts and citrus
- ❑ cereals—cereal crops occasionally can suffer from frost and only rarely do cereal farmers insure for frost
- ❑ grapes
- ❑ flowers
- ❑ vegetables.

Other frost strategies include:

- ❑ picking least-risk sites to plant crops or orchards. Frost usually occurs at the lowest point of a farm, where air drainage is poor, so susceptible crops should not be planted in such areas.
- ❑ preparing the ground properly is also important to help prevent frost damage. In grape vines, this may mean keeping the ground flat, clear of weeds and sometimes rolling the soil—all these factors help air drainage, allowing cold air to flow off the area and away to a lower site where it will do less damage.
- ❑ later sowing in cereals encourages later flowering—this may move cereal flowering outside the season's high-risk period.
- ❑ using overhead irrigation to apply water to flowers and buds—in the event of frost the water may protect tender new growth from frosts. Some farmers also put frost-retardant chemicals in irrigation water which also helps protect the plants from frost damage.
- ❑ delaying pruning may help delay flowering and fruit-set in some crops, so avoiding frosty periods.
- ❑ frost fans are used in some vineyards to circulate air and prevent frost settling on grapes.
- ❑ field burners are occasionally used. These burners use sump oil as fuel and produce smoke which blankets the ground, producing an

insulating layer of smoke particles that helps prevent frost settling on the vines.

Practical aspects of heat risk

Like cold, heat can cause operational problems and therefore create physical and financial risks for farmers. High temperatures require a range of measures to minimise the risk they pose to farming operations.

Some risks high temperatures pose to farming include endangering the lives and/or growth rates of:

- ❑ small animals, such as chicks, bees, piglets and lambs
- ❑ animals that are intensively housed
- ❑ animals that cannot get access to either shade or adequate supplies of good quality drinking water.

In the late 1980s large numbers of laying hens near Cairns had to be put down because a tropical cyclone destroyed the infrastructure of many farms, destroying supplies of drinking water. Unfortunately there was no relief available to the poultry from heat and thirst, so farmers and the Queensland Department of Primary Industries staff had no choice but to destroy the birds.

Plants also can suffer in hot conditions. For example, they can:

- ❑ suffer from water stress
- ❑ be more susceptible to certain diseases and fungi.

Farmers can minimise risks by:

- ❑ providing cooling for at-risk animals
- ❑ insulating sheds
- ❑ providing shelter and shade for free range stock
- ❑ ensuring crops have adequate water where possible.

Wind

Dealing with wind is a normal part life on the land, but excessive wind poses risks. Too much wind can cause problems that may include:

- ❑ physical damage to improved structures, crops and natural vegetation
- ❑ heat stress
- ❑ cold stress.

In southern Australia, wind storms and hail may occur together, and both can cause damage. Hail is less common in the tropics, but it can occur. Rain may also worsen the effects of wind and cause significant stock losses. Farmers can provide livestock with shelters to cut such losses, particularly with newly-shorn sheep, new-born calves and lambs.

Wind may severely damage unprotected soils by scouring away the topmost layers of the land. This is why many cereal farmers now practice minimum tillage, to protect soils from the scouring effects of wind and rain.

Practical aspects of wind risk

Wind damage can be reduced by using wind breaks in horticultural plantings. Other strategies include wind damage insurance, which is often available in a combination with hail damage insurance. It can be obtained for the following crops:

- vegetables (a combination of wind and hail damage insurance)
- tree fruits (separate wind insurance cover can be taken out)
- cereals (a combination)
- vines (a combination)
- avocados (a combination)
- nuts (a combination)
- tropical tree crops (a combination)
- it is also possible to insure against losing trees due to wind.

Fire

Fire is a common event in Australia. When fire occurs on farms it can have devastating consequences. Fires can destroy crops, fodder supplies, livestock, property and people, so minimising the risk of fire is a very important task for farm managers.

Strategies to prevent fires are well documented. This information is readily available from places like local councils, volunteer bushfire brigades, and appropriate government departments.

There are a number of strategies farmers can use to cut fire risks. These include:

- making fire breaks
- keeping effective fire-fighting facilities on the property
- becoming a part of local volunteer fire-fighting network
- maintaining a sound and effective communications system.

Most farmers keep fire insurance to cover against the financial losses caused by fire. This insurance can be obtained for most farm items, including:

- crops
- livestock
- farm improvements, including fences, sheds, houses
- life, as life insurance.

It is also possible to take out insurance to cover a loss of business income. This means your property can still maintain a cash flow. That means your viability will not be threatened if a fire damages part or all of your business.

Even small fires can have a drastic effect on operations

Flood

Like fire, flood also can have a devastating effect on a property. Floods can damage crops, kill livestock, damage machinery and fixed improvements plus degrade land.

Managing floods is difficult as water, like fire, can be unpredictable. The 1992 flood in the mid-west of NSW is a case in point. In that incident, a large area of farming land and the town of Nyngan was flooded. There was little anyone could do other than carry out rescue work.

In flood-prone areas a disaster plan is very helpful, allowing rescue plans to be acted on without delay. On individual farms there may be some things you can do to minimise the effects of floods. For example, you may have your own plan of action to put into effect if flooding looks imminent. Unfortunately, in large, flat areas like western NSW, these on-farm anti-flood options may be fairly limited.

Practical aspects of flood risk

Managing flood risk over large areas is really a community problem and one where governments have a major role. Flood mitigation schemes and dams can sometimes be used to reduce the risk of floods, but these are expensive.

Individual properties can use levee banks to control flooding, but a property's location has the greatest influence on its flood risk.

If flood prone areas are uninsureable, establish important assets on more promising land

There are two main types of floods:

- ❑ floods from over-flowing watercourses—which can cause widespread flooding
- ❑ floods from water ponding—which causes local flooding.

Over-flowing watercourses cause the most common forms of flooding in Australia. This type is the one form most difficult to manage. It is often difficult to get insurance for this type of flooding, due to the concentration of high risk properties near watercourses.

The second form of flooding is due to excessive rainfall on properties. Insurance cover for this type of flooding is easier to get than watercourse flood insurance.

4 Drought

Drought is a common occurrence in Australia. It is easy to describe but quite a bit harder to define. This difficulty of defining drought happens because:

- ❑ no two droughts are alike
- ❑ droughts can have varying influences in different agricultural districts.

For example, a high rainfall area can show green pastures to a casual observer, yet farmers can be experiencing drought conditions. Below average rainfall means normal grazing pressures cannot be maintained on those pastures. This situation is quite a common effect of drought in Tasmania and tropical north Queensland. Sugar cane too can look green but be drought-affected. Drought, therefore does not necessarily mean bare paddocks and blowing dust. In most areas of Australia though, a drought does result in barren land, poor crops and starving stock.

The financial influence of drought is of most concern to farmers. Stock can either be fed supplements or be moved away for agistment. Crop plantings too, can be held over or delayed, but these strategies all cost money. Therefore the biggest business impact of drought on a property usually depends on:

- ❑ financial resources
- ❑ the strategies in place to cope with the financial demands drought will place on business operations.

Drought and politics

The issue of drought has received a lot of attention by both Federal and State Governments over recent years. Governments now are less keen to provide funds for drought relief.

Governments are increasingly seeing drought as a natural climatic characteristic of the Australian environment. Their argument is that drought should be managed by farmers as a part of farming's normal financial management. Governments argue good managers have coped with drought using their own resources, while others have sought government drought relief. Some governments claim the present system of government drought relief is unfair and encourages poor drought management planning.

There are two main reasons for current government attitude to drought. In previous droughts, governments provided grant money to farmers in the form of fodder subsidies, agistment subsidies and livestock transport grants. Government, both bureaucrats and politicians, concluded those farmers who mis-managed their financial resources were then eligible for drought subsidies. On the other hand, those farmers who managed well were not eligible because they were not seen to be in financial need. As a result, the system of giving farmers drought grants favoured poorer managers. Hence, governments felt the drought grant system was a disincentive to those farmers who tried to plan for drought.

Secondly, in recent times, State and Federal Governments have been trying to cut budget deficits. Because of this, there is now strong financial pressure for governments to cut drought relief payments.

The results of these changes in government policy mean:

- ❑ there will be more pressure on farmers to be totally self-reliant through their own management
- ❑ government assistance may be totally withdrawn, except in unusual circumstances like the recent 1991/94 drought in Queensland—that drought was so severe and so prolonged that, without outside assistance, few farmers could have predicted or coped with a drought on that scale.

Your drought strategy will obviously depend on your own opinion about continued government drought assistance—and what the result of the current drought debate will be.

Defining drought

Usually we think of drought as a growing season with minimal and perhaps no useful rain, because drought is usually measured by lack of average rainfall.

Figure 4.1 illustrates the rainfall patterns and the major droughts affecting Australia in 1902, 1928, 1946, 1965 and 1982.

A recent Federal Government inquiry challenges this view by introducing the concept of relative drought. Here, the seasonal conditions and agricultural production requirements often do not occur together. In addition, drought strategies clearly need to be different at high stocking rates than at

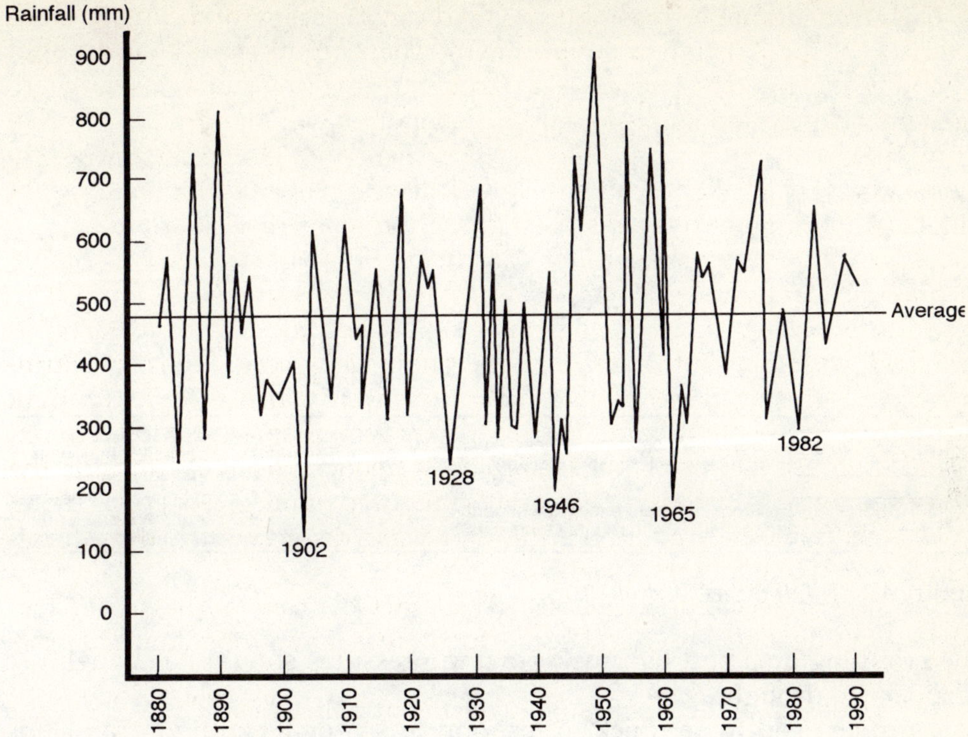

Figure 4.1 *Traditional view of drought*

low rates under the same pasture conditions. For a heavily stocked property and a lightly stocked one in the same district, drought is a relative term. In the same district, one property may be short of feed while the other will still have feed to spare.

The idea of 'relative drought' is an important concept to understand and incorporate into your farm business drought strategy.

To take the idea of relative drought to its logical conclusion, there should be little risk from drought, as the Federal Government says farm managers should implement management strategies to physically and financially handle any drought successfully. Unfortunately big droughts like the 1991/94 Queensland drought proved the theorists wrong. Almost no-one on the Darling Downs could have coped with that drought alone—no matter how good they were as managers.

A recent Federal Government inquiry to assist in setting a national drought policy defined drought as follows:

> Drought represents the risk that existing agricultural activity may not be sustained, given spatial and temporal variations in rainfall and other climatic conditions.
>
> [Spatial means from place to place and temporal, from time to time.]

The key to successfully managing your property's drought situation is to closely match the seasonal conditions to your farming operations. Hence

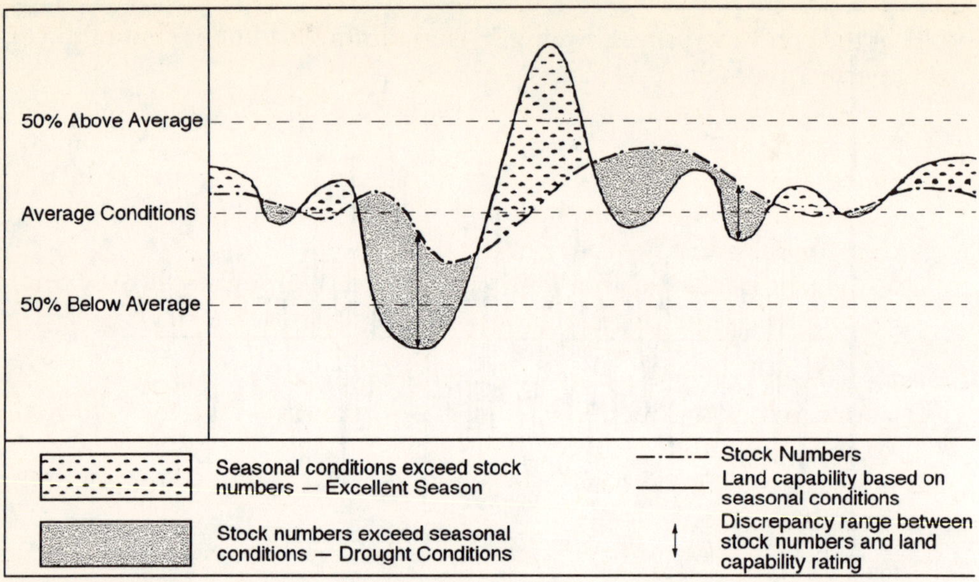

Figure 4.2 *Relative view of drought*

the cropping area and livestock grazing pressures should be changed in response to drought.

Farmers need to be quick to respond to seasonal conditions and fine tune their property's productivity accordingly. Your drought strategy should include climatic 'trigger points' that when reached trigger either a drought strategy or a good season strategy.

Examples may be:

- ❑ if it hasn't rained by May 31, start dry seeding a third of your normal cropping area
- ❑ if there has been no monsoonal break to the cattle season by February 10, continue feeding urea and molasses, but as a drought supplement, not a normal seasonal supplement.

Recognising drought

The major problem in managing the effects of droughts is recognising that a drought has actually started. Most producers need a defined trigger point at which they will start to implement their drought strategy. This strategy and its trigger point need to be clearly thought out and defined prior to any drought occurring.

Here are some examples of drought strategy triggers based on the farming enterprise.

Dryland-cropping drought triggers to start your drought strategy may include the following points:

- ❑ if the combined pre-growing season rainfall is ten per cent of the expected average
- ❑ (in the cereal belt) if there is no usable rainfall by your set date, start dry-seeding
- ❑ if less than 30 per cent of average rainfall was received in the opening wet months (April, May and June in southern Australia and December, January and February in the tropics)
- ❑ if the long-range weather forecaster predicts a drought for your farm—if you believe in long-range weather forecasts.

Livestock drought triggers may include the following:

- ❑ if dry matter production is 30 per cent below average for your main growing months
- ❑ if no usable rainfall has been recorded for the first four months of the growing season
- ❑ if only 30 per cent of average expected rainfall has been experienced for the main feed-growing months.

Pastoral drought triggers may include the following:

- ❑ if no useful rain has been received for twelve months
- ❑ if your property received two consecutive years of low rainfall, and less than half of the average expected rainfall.

Practical aspects of handling drought risks

Practical drought management starts well before drought occurs.

It needs clear, logical thinking to put a drought strategy together—so it is vital to have a range of options to choose from, especially if you are going to make effective preparations before a drought begins.

Make extensive use of your 'what if' budgets (or partial budgets) while you are putting your drought strategy together. Once you have formed your strategy, your biggest decision will be to decide when a drought has actually started.

You should try to work out your drought trigger points while droughts are not imminent, so your decision remains logical and without emotional prejudices.

Having formulated your drought strategy, don't be tempted to change it during the drought without very good reasons—your decisions during the stress of a severe drought may not be too sound.

Cropping

Drought management strategies designed to cut risks may include the following:

Sowing early with minimal opening rains
Dry seeding, or sowing with minimal opening rains, are options when early rains are inadequate.

Growing days
Calculate the variable costs and break-even yields. Then work out the growing days required to grow a financially successful crop, given the expected rainfall for the rest of the growing season.

Monitor the rainfall deciles to guide planting decisions
Monitoring allows people to make judgements about the probability of growing season rainfall. This technique can be tailored for a particular property and provides useful extra information to help you decide when to activate a drought strategy.

Review cashflow to decide added financial requirements
Continually review your cashflow budget, so you can keep abreast of your cash position. This will allow you to negotiate with your bank for funds you may need for the following season.

Keep in regular touch with your banker and show the bank your drought cashflow forecasts early on in the season. This action will improve the chance of your bank manager better understanding your position and therefore giving you the assistance you need.

Cropping strategy once the drought has broken
Once a drought has broken, increase the cropping area for the next year if stock numbers are down. Because of reduced cropping during the drought, less nutrients will have been taken from the soil. This is a benefit, but you should review your fertiliser levels to maximise your returns in the improved conditions.

Broaden your income base
If possible, try to develop some off farm investment or income portfolio so that income may still come into the business during poor farm income years.

Livestock

Drought management strategies designed to cut risks for livestock enterprises may include the following

Agistment
This strategy ensures adequate feed and water supplies are available to stock because of their move to non-droughted areas. However moving stock can increase the risks of disease.

Where sheep and other fibre animals such as angora goats and alpacas are involved, you also may have to make provision for shearing.

Keeping feed reserves on hand
It is usually cheaper to produce and store your own drought feed reserves than buying in feed during a drought. This strategy will require careful budgeting, as it costs money to handle and store fodder. For example, there are interest costs in having money tied up in feed, silos and sheds.

In a drought, feed is often scarce and expensive, so if you reach a trigger point, it may be wise to think about including a fodder purchase scheme in your drought management plan. That way you will be able to get feed on hand at fairly low prices. Any surplus feed you have may also be sold at a good price to help offset what the drought is costing your business.

Buying feed
Buying feed to maintain your core breeding animals is a strategy you should consider for your farm. However, as in the point above, buying in early when your drought plan is triggered may be a better strategy than buying expensive feed at the height of the drought.

Selling stock
Another drought option is to identify your core breeding stock and keep these. Then cull all surplus or less productive stock before the drought hits hard and stock prices crash. It still may be financially sensible to sell surplus stock at drought affected prices to minimise losses and reduce the feeding costs.

Drought may necessitate the use of more resistant breeds or grazing the 'long paddock'

Using roadsides
As a last resort cattle and sheep can be driven along roadsides in search of feed.

Combining strategies
One management option might be to use combinations of the above strategies. This may depend on the length of the drought. Each part of the strategy can be implemented as the drought continues.

The importance of body weight measurements

Keeping track of livestock weights may indicate how well livestock are holding up to drought feeding. For example, with sheep, weights can be monitored weekly by either randomly weighing 10 per cent or 100 sheep out of the centre of a mob, or tagging 30 out of 500 sheep. Weighing is best done at the same time of day to allow for the variation in gut fill. This method can be adapted for other stock, including cattle.

Another way of monitoring sheep is by not letting them fall below a fat score of two. (See your Department of Agriculture or Primary Industries for more information on drought feed rations and fat scoring.) Similar information is also available for cattle in some states.

The importance of energy
Food units, also called starch equivalents (SE's) are used to measure feed energy. Food units are a vitally important measuring tool in drought feeding stock.

Four SE's per week are usually adequate for adult sheep. Vitamins will rarely be required especially if there is a green pick from time to time. Weaners may need vitamins if they are on a wholegrain ration. Where lucerne or other good quality hay is used, no vitamins will be required.

Three SE's are adequate for weaners, but on grain the weaners will need a protein supplement, such as sheep nuts, lucerne pellets or lucerne hay.

Normally cattle need about ten times the nutrients of sheep, so yearlings would need about 30 SE's using this ten times rule since weaner sheep need three.

Protein is also important with weaners, calves, lactating ewes and cows. Calcium or powdered limestone at 1.5 per cent should be added to rations when feeding wheat, as wheat is low in calcium and may not supply enough to meet the needs of young, growing stock or lactating older animals. Phosphorus is another important feed additive in phosphorus deficient north Australian pastoral country. Phosphorus deficient cattle will often chew carcass bones as they seek out phosphorus. Bone chewing is a highly dangerous practice, as this is the major cause of botulism, one of the biggest cattle killers in the dry, phosphorus deficient areas of northern Australia.

Good supplies of clean water

Good supplies of clean water are very important for stock horses, weaners, calves and lactating sheep and cattle.

Water intake increases with urea feeding and water storages will often become more saline as a drought lengthens—so watch water quality.

If water supplies run out farmers can often have a 'water drought' before a 'feed drought' as happened near Mt Isa in 1991, though the prevalence of bores meant the water drought never became a big problem.

Crop stubbles

In cropping areas, crop stubbles can keep stock alive, especially if stubble feed is supplemented with either a protein supplement or urea.

Which feed to use?

Minimising feed costs is important and this should be done by comparing the cost per SE of different feed sources. Table 4.1 shows typical starch equivalents, or feed value per kilogram, of some common drought supplementary feeds. From this table you should be able to calculate the cost per SE.

Table 4.1 *Feed values*

Feed Types	*SE/kg*
wheat, barley	1.54
oats	1.32
sheep nuts	1.43
good lucerne hay	0.88
good legume and cereal hay	0.79
poor hay	0.55
barley and oat straw	0.44
wheat straw	0.40
silage	0.18

For low quality feed, sheep and some cattle may not be able to eat enough material because of limited gut capacity. As animals may die under even slight stress, you may prefer to use a more expensive but safer feed.

It is also sound practice to check that correct quantities of feed ordered have been delivered to the property.

Calculating the cash cost per SE

For this example we will assume the costs are costs per SE delivered to the property, not costs at the source.

Barley contains 1.54 SE/kg. If barley is priced at $1.30/kg then the food value is priced at $0.84/SE (that is $1.30 divided by 1.54). This is the calculation needed to decide the best feed prices per available SE.

Very good lucerne hay would be 1 SE/kg. Early cut hay with clover in it would be 0.78 SE/kg, poor hay would be 0.67 SE/kg and poor wheaten hay would be 0.44 SE/kg.

Break-even feed costs

From the calculation above, find the cheapest SE/kg cost and decide the break-even prices that the other feeds have to achieve before you would select them for a drought ration.

For example, if barley was the cheapest at $0.84/SE and the costs of oats was $140 per tonne, you would work out the break-even price for oats as follows:

SE cost for oats per kilogram is 0.14 (cost/kg) ÷ 1.32 (from Table 4.1) = 0.11 per SE.

For oats to equate to barley the following calculation is used:

$$\frac{\text{Price of oats}}{1.32\ \text{SE / kg}} = \frac{0.14\ \text{(price of barley)}}{1.54\ \text{SE / kg (barley)}}$$

$$\text{Price of oats} = \frac{1.32 \times 0.14}{1.54}$$

$$= \$0.12\ /\ \text{kg}$$

$$= \$120\ /\ \text{tonne}$$

So the price of oats would need to fall to $120 per tonne before oats would be a better option than barley as a drought ration.

How much feed do stock need?

Table 4.2 provides a list of SE values for several types of feed. You should also collect similar information for feeds used in your area. This table deals with sheep. Cattle need about ten times the feed SE's per head per day as do sheep, so for cattle producers, increase these values by ten to get an approximation for your stock.

Confining stock

There are several advantages in confining stock during a drought:

- ❑ confining gives better control of both quality and quantity of food eaten
- ❑ it takes less work to feed confined stock
- ❑ confining stock allows pastures to grow faster when the drought breaks.

Feeding frequency

Frequent feeding is best when starting a drought ration.

Feed once a day or every second day to begin with, then graduate to twice weekly and finally organise weekly feeding if your vet advises.

Table 4.2 *Drought maintenance rations per 100 sheep. (Assume weekly requirement is 4 SE per sheep.)*

Feed	***Entering Drought***	***Some Roughage Available***	***Complete Hand Feeding***	***(6 months)***
Wheat & Barley	63 kg	126 kg	252 kg	6.4 t
Oats	80 kg	159 kg	318 kg	8.1 t
Good hay	120 kg	239 kg	479 kg	12.2 t
Sheep nuts	70 kg	140 kg	280 kg	7.1 t

When planning infrequent feeding, cost versus animal survival are important factors to take into account.

Introducing grain rations

Many stock are fed grain rations in southern Australia because of the ease of transport and the relative concentration of food in grain compared to hay. Grain should be introduced gradually and can be done by mixing rations and increasing the proportion of grain to the required level.

Early weaning

This is a very sound practice, as it reduces stress on mothers, but young animals must then be fed on demand with good quality rations.

When the drought breaks

Stock prices usually don't increase immediately after a drought has broken. This is partly because people are not convinced the drought has ended, and because there isn't usually much money about for restocking.

When rain comes and if it is accompanied by cold weather, it may be necessary to increase rations for a while. Often the biggest stock losses occur when drought breaking rains start.

Broaden the property's income base

Try to develop an investment portfolio or income source independant of the farm to ensure some income is coming into the business during poor farm income years.

Action

Having considered the above, decide what to sell early—old stock, dry ewes and cows, cull animals and so on.

Decide what feeding strategy you will adopt for the stock you keep, and revise your budgets.

Decide too if you think you may need more finance. If you do need more money, go and see the bank or stock firm before everyone else gets there.

Pastoral

Pastoral drought strategies have some similarities to the livestock strategies outlined above, so drought management strategies designed to cut risks may include the following.

Agistment

Transport stock to other properties in areas unaffected by drought and agist stock while the drought lasts. It is important to take care with stock health as animals can suffer from additional health problems in new locations.

Efficient feeding

Buying feed during droughts can be expensive. There are special feed porridges that can be made which provide essential proteins and minerals. Information on these porridges can be obtained from your local Department of Agriculture or Primary Industries.

Graze stock on stocking routes

A common practice is for stock to be driven along stock routes to graze the available feed. Transporting stock to these routes outside the drought area may be an option.

Culling stock

Cull and sell less productive stock early in the drought. This will allow you to feed the nucleus of the flock or herd.

Good supplies of clean water

Ensure water supplies are well maintained and are of good quality.

Confining stock and breeding control

Controlled breeding gives many producers an option of not mating stock through a drought. Confining stock during drought also gives the following advantages:

- ❑ reduces soil erosion
- ❑ assists in easy drought feeding; and
- ❑ makes it easier to manage stock.

Early weaning

Early weaning of livestock gives lactating animals a greater chance of surviving the drought.

Animal health

Managing animal health is more important in drought conditions, as animals are more susceptible due to their weakened condition.

Pest control

Pest and vermin control, particularly of pigs, rabbits, donkeys and goats helps cut feed competition. Controlling predators like dingoes and foxes also may stop weaker animals becoming prey.

Fodder trees

If available, fodder trees such as kurrajongs, willows and tagasaste are a good source of stock feed in dry times.

Broaden the income base

Having off farm investments that can generate income also helps keep a cash flow going through droughts.

5 Short term commodity price fluctuations

Commodity price fluctuations have been, and will continue to be, prominent characteristics of Australian agriculture. As shown in Chapter 1, commodity price fluctuations are also one of the more significant influences on farm profitability. If the risk caused by commodity prices can be reduced, then a large part of farm risk may be overcome.

Most of Australia's rural commodities are exported. Therefore the state of international economies and markets have an important bearing on prices, as these affect commodity prices received by Australian farmers. Continually monitoring and improving knowledge of these overseas markets is becoming an increasingly important task for farm managers.

The international economy

The international economy can influence farm gate prices for commodities in three main ways:

- exchange rate levels
- trade restrictions
- international commodity price movements.

Unfortunately, both trade restrictions and exchange rate movements cannot be influenced by individual farmers. While commodity price movements are also largely beyond farmer control, there are several management tools and techniques that are useful in managing their associated risks.

Exchange rates

Most of Australia's exported commodities are traded in the currencies of other countries. As the US is the world's largest economy and is often the biggest international trader, most commodities are traded in US dollars ($US), even though a particular commodity may never enter the US market. Hence, any shifts the Australian dollar ($A) makes against other currencies will have a direct impact on prices within Australia.

For an example, say the price for peas is $A200 per tonne and the $A/$US exchange rate is $US0.70 ($A1.00 = $US0.70). If the $A/$US exchange rate fell tomorrow to $US0.69, the new $A price would be $202.90 per tonne (200 x $0.70/$0.69).

The rule on exchange rates is:

If the $A increases against the $US it is more difficult to export but easier to import into Australia.

Conversely, if the $A decreases against the $US it is easier to export but more difficult to import into Australia.

If the $A increases against the $US, farmers find their export prices less competitive and therefore it is harder to sell products, while imports, such as electrical goods and cars become cheaper.

Exchange rate movements can therefore have a big impact on farm gate prices.

Trade restrictions

The current agricultural trade war between the European Community (EC) and the United States (US) is a good example of trade restrictions. Some of the world's cereal markets have artificially low prices because both the EC and the US are supplying these markets with subsidised grain. While these markets do not constitute all the world's markets, this activity does depress world cereal prices.

Other forms of trade restrictions include tariff barriers, import quotas and complete restrictions on imports.

International commodity price movements

The demand and supply of agricultural produce is continually changing, due mainly to the variation of supply from season to season. The main influence on supply is world climate. Supply is one of the main influences causing international price uncertainty, and hence price risks to Australian farmers.

Figure 5.1 shows the world price for wheat ($A) over the last few years. You will notice it fluctuates quite widely, with no reasonably predictable pattern. The large world price fall in early 1990 was brought about by record wheat production in the EC and high production in the US. This increased supply caused wheat prices to fall.

Wool prices are also dependent on international demand, which has been greatly reduced over recent years. Wool prices have been decreasing due to

Trade wars between the US and Europe can affect commodity prices across the board

the huge oversupply in world markets. This oversupply has been due to:

- ❑ Australia's large wool stockpile
- ❑ Australia's relatively high annual wool production
- ❑ Australia's dismantling of its wool floor price.

As a result of the depressed international demand for wool, prices are expected to remain low for several years.

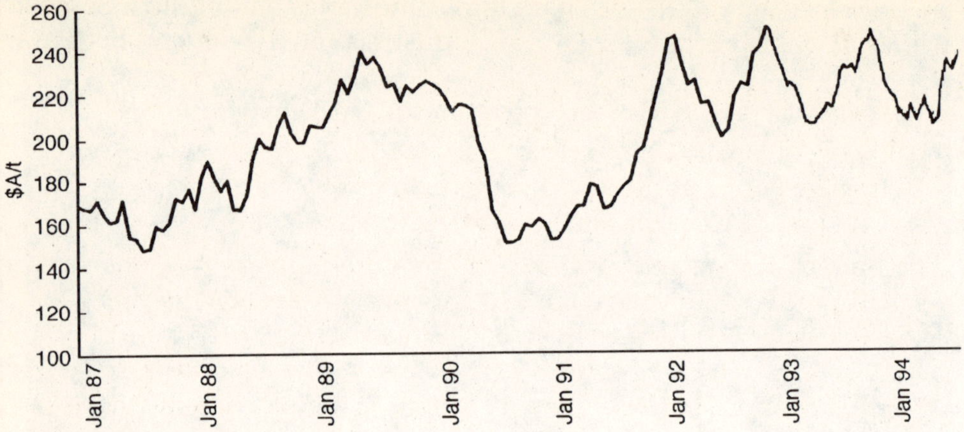

Figure 5.1 *World wheat price*

Nevertheless there has been some sign that wool prices are picking up, because western European and Chinese wool buyers have recently started returning to the market.

The domestic economy

The markets that your farm products are being sold into will decide whether the domestic or international economies will have the greatest influence on your property's financial performance.

Although most Australian rural produce is exported, a few industries rely heavily on the domestic market. These include most fruit and veg-

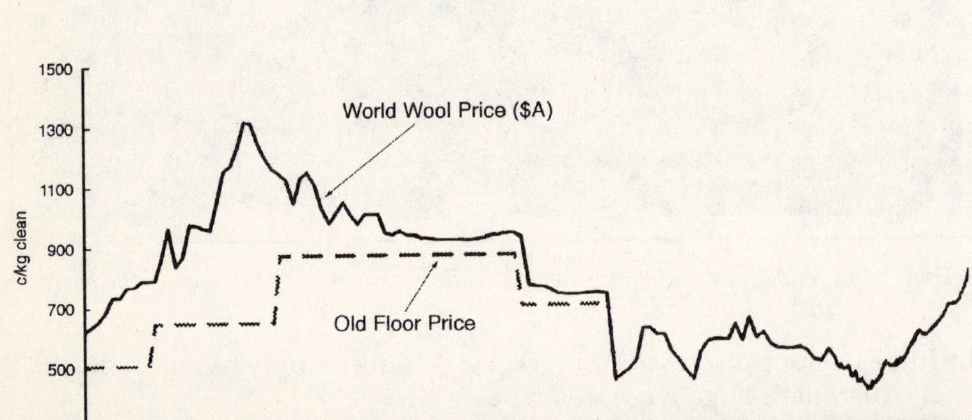

Figure 5.2 *World wool price*

Trade wars between the US and Europe can affect commodity prices across the board

the huge oversupply in world markets. This oversupply has been due to:

- ❑ Australia's large wool stockpile
- ❑ Australia's relatively high annual wool production
- ❑ Australia's dismantling of its wool floor price.

As a result of the depressed international demand for wool, prices are expected to remain low for several years.

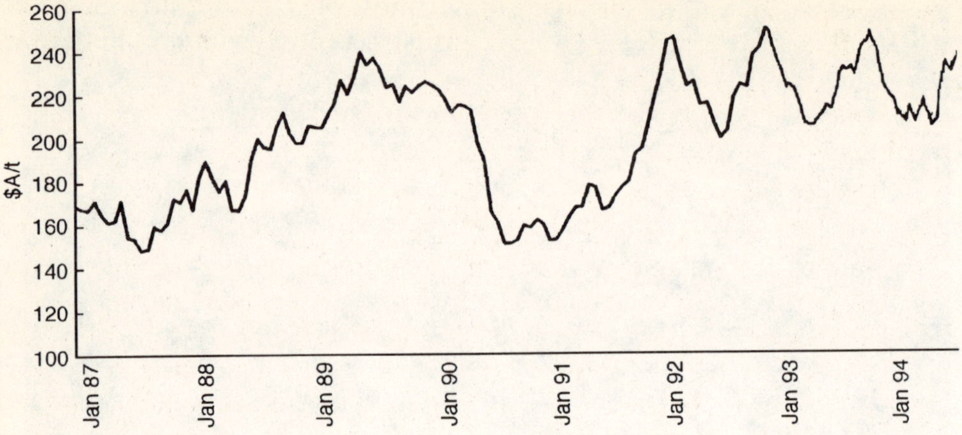

Figure 5.1 *World wheat price*

Nevertheless there has been some sign that wool prices are picking up, because western European and Chinese wool buyers have recently started returning to the market.

The domestic economy

The markets that your farm products are being sold into will decide whether the domestic or international economies will have the greatest influence on your property's financial performance.

Although most Australian rural produce is exported, a few industries rely heavily on the domestic market. These include most fruit and veg-

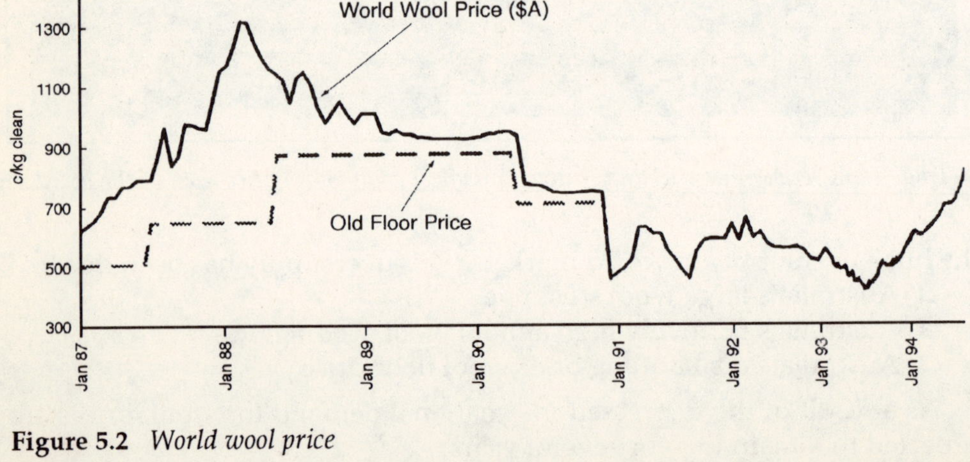

Figure 5.2 *World wool price*

etables, wine (though this is changing), lambs, pigs and poultry. Some industries such as the dairy and meat industries rely heavily on both markets.

The main factors that affects the domestic market are:

- the level of supply and demand
- the level of protection various industries receive
- to a lesser extent the state of the Australian economy.

Vegetables are a classic example of a mature industry that depends on seasonal conditions in various producing regions around Australia.

Due to excellent transport most regions can market their produce in any major vegetable market in Australia. Therefore vegetable growers often only make good profits when one of their competing regions suffers yield losses, through flood, hail, frost or too much rain at the wrong time.

Commodity price forecasting

It is very difficult to accurately forecast commodity prices twelve months in advance. It is even harder to predict these two to five years ahead. However, if you can accurately predict the direction of price movements, then farm plans and risk strategies can be put in place to either maximise farm profits or minimise losses depending on whether you expect buoyant or poor market conditions.

Some farmers, and bank managers for that matter, rely too heavily on the prices of last season to guide budgeting for the coming season. With increased price variability, brought on in some industries through deregulation, this forecasting method is becoming less accurate.

The main steps in assessing commodity price movements are :

- knowing who are the major suppliers and consumers
- understanding the current supply and demand situation
- monitoring the planting area and seasonal influences of your major competitors (or have a reliable information source to assist you)
- knowing the quality of product you produce and what premium it is likely to bring
- understanding the developing stock situation (how much of that commodity stock is being held around the world)
- understanding the influence any current trade restrictions have in your market
- appreciating the trading policies of any marketing authorities you use, such as the Australian Wheat Board
- assessing other market indicators for their expectations of future price expectations.

Practical aspects of handling short term commodity price risks

Strategies designed to manage price risks will differ between farm managers—depending on farm enterprises and the risks each manager is willing to take. The old activity of managing the farm the way 'Dad did it' is fast changing as the structure of markets changes dramatically. Now, managers need to think much more about marketing their produce than they did in the past.

There are strategies available to manage the risks of commodity price movements, but these vary between industries. The main tools available include:

- using the futures market
- using forward physical contracts
- using the spot market (the current market price)
- diversifying the farm enterprise.

The futures market

Futures markets allow people to buy and sell futures contracts. These futures contracts are contracts for sellers to deliver commodities to buyers at a future date. Contracts are commonly written for wheat, cattle, orange juice, sugar, cotton, gold, silver and so on. Futures markets are different to traditional markets, because a futures market is a market where risks rather than commodities are traded. Most farmers use futures markets as a form of price insurance for commodities.

A producer can enter into a contract to sell some steers to a meatworks in a year's time for say $1.50 per kilogram. If the price goes up to $2.00 then the producer will lose $0.50 per kilogram. If the price drops to $1.00, then the producer makes $0.50 per kilogram. It may be that getting $1.50 per kilogram for steers is good enough for the producer, so taking out a futures contract guarantees an income of $1.50 per kilogram, regardless of what the real price is in a year's time. In effect this producer's futures contract has become a form of price insurance.

Futures markets are often very complicated to use, so you should use a futures broker to help you. If you use the market purely for price insurance purposes (hedging) and not to speculate to make money, you should only lose the equivalent of an insurance premium. However, if you speculate, you have the potential to make quick profits or quick losses—and the losses can be substantial. Such speculatory losses have led people to believe the futures market is a high risk market. If futures are used only as a means of price insurance, there is no risk of losing any more than the risk premium. This is very similar to fire insurance, where a year of no fires means you still have your sheds and dwellings, and the insurance company keeps the premium for that year.

Forward physical markets

Selling forward physically means a supplier and buyer usually make a contract for:

- ❑ a fixed amount of commodity
- ❑ at an agreed price
- ❑ at a specified quality
- ❑ to be delivered at a certain time.

An example of forward selling are the contracts offered by the San Remo company. San Remo makes pasta and it needs regular supplies of durum wheat with a certain protein level for its product. San Remo therefore offers farmers contracts (forward physical contracts) to grow and supply their durum wheat.

This type of contract is also available from G H Michel Ltd for wool producers.

The spot market

The spot market is the term used in the futures market to mean current market prices. Most farmers use current market prices at the time of sale to decide their entire farm income. For example, a farmer has little control of the world price. The only control is to place a reserve on produce if it goes to auction. Grain legumes are sold at prevailing harvest prices and prices for export steers are set after meat processors do their sums based on what the US or Japanese beef markets are offering.

Using spot markets means a farmer is at the mercy of market price rises and is never certain what the final price will be. This approach can often be more risky than using some of the price insurance mechanisms already mentioned.

Diversifying the farm enterprise

This strategy of spreading risks has been successfully practised by farmers for years. For example a grain grower may produce wheat, barley, oats, peas and vetch, which helps spread risks. Not all grains are affected by climate equally and their markets react relatively independently. This means while one market is low others may be high, so helping maintain a reasonable farm income.

It is possible to also spread risks between completely different enterprises, such as wheat, sheep and cattle. Usually cereal, beef and wool price cycles are not synchronised, so diversifying helps stabilise farm incomes.

Another example of spreading risks is to develop off farm investments. Farmers who develop sizeable investments off farm have a natural form of income insurance. For this to occur, it is important to maintain a long term plan where not all farm profits are put back into the farm. This system tends to work well, as domestic cycles often run differently to international commodity cycles. Therefore, while farm commodity prices may mean lower farm income, the off farm investments may well be giving improved returns.

Combined strategies

A strategy of employing parts of all the above strategies is an ideal way of managing selling risks. These strategies may not mean people will always get the highest prices—but by spreading risks, there will be less danger of getting very low prices. This results in a more stable income in place of fluctuating high and low prices.

For example this strategy may result in a farmer putting a third of a crop on the futures market. If the insured price appears attractive, one third could go in a forward physical contract and the rest could be sold at prices prevailing at the end of the season.

6 Long term commodity price fluctuations

When we think of commodity price movements it is common to think of prices that can move quite freely between seasons. Examples are wheat, grain legumes and boneless beef. The price movements of these commodities have the most dramatic effect on cashflow and so are most noticeable.

Long term movements in prices are not often considered, as they are less noticeable. However long term price movements are one of the main reasons the rural community is finding economic survival so difficult. This economic difficulty affects both individual farm businesses and whole rural communities.

These long term shifts can creep up on rural businesses very slowly and create increased risks to farm financial viability. Managing these risks is vital for your farm's long term viability.

Evidence of long term change

The most dramatic evidence of a long term shift is shown in the Farm Terms of Trade Index (Figure 6.1).

The Farm Terms of Trade Index measures the movement of prices received and paid by farmers. If the index moves down it means prices paid by farmers have increased more than the prices they received.

Since the early 1950s, the Farm Terms of Trade Index has been declining. Although there have been periods of rural prosperity during the periods from 1972 to 1974, 1979 to 1981, and again during the years of high wool prices from 1988 to 1990, the long term trend has been down.

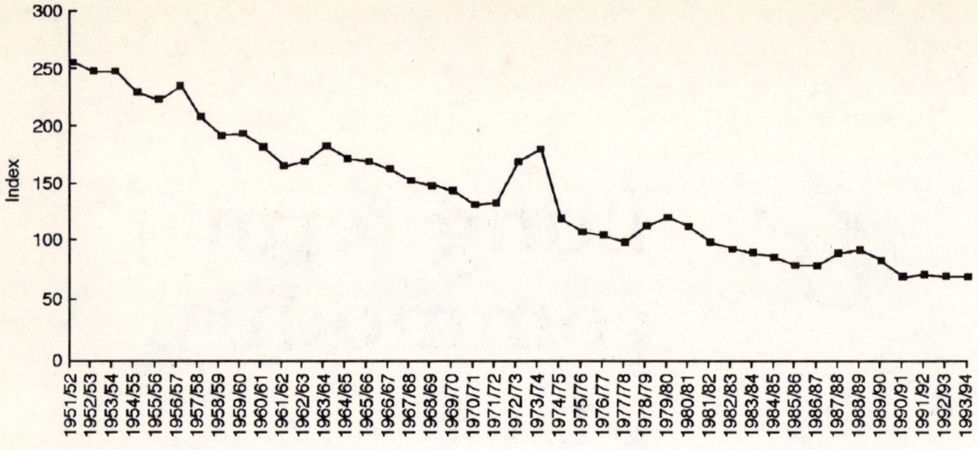

Figure 6.1 *Farm terms of trade*

It does not allow for:

- ❑ improvements in productivity brought on by improved technology
- ❑ improvements in efficiency that could occur with farm amalgamation
- ❑ cost cutting on individual farms.

A continual downward trend in the terms of trade is not necessarily evidence that long term prices have changed, as prices could have remained unchanged while prices paid have been increasing. However a falling Farm Terms of Trade Index is one indication that farmers are receiving constant real cuts in their profits.

Farm terms of trade

Will farm terms of trade continue to decline? It is vital each farm manager forms an opinion and incorporates these ideas into any long term plans for the business.

Economic theories indicate declining terms of trade will continue in Australia. This is characteristic of developed countries like ours. The US and EC also have been experiencing this trend, which is one of the reasons for the trade war between those trading blocs. Farmers in those countries have a lot of political influence and have therefore obtained government assistance to combat the effects of their own declining terms of trade.

There are a few exceptions to the trend which may provide medium term relief for Australian farmers. These are:

- ❑ the continual economic development of the Asian countries
- ❑ the initial import requirements of the emerging east European countries.

Asian countries

As developing countries improve their living standards, people's diet changes from subsistence foods, based on cereals and legumes, to a more Western diet. Therefore demand for foods such as breads, fruits, vegetables, dairy products and meats increases.

The initial policies of many Asian countries is to be self-sufficient in this type of food production, which does not encourage imports. However as the standard of living increases the demand for these western foods outstrips the ability of the country to match supply, so imports are brought in to satisfy demand.

As Australia is a key world exporter of these foods, our future food exports to these countries appears quite bright. Therefore the declining trend of farm terms of trade may be slowed or even stopped in the medium term.

Eastern European countries

The break up of the USSR means that once the newly emerging countries have stable governments, they should begin to achieve improved economic growth.

With economic growth comes improved living standards and increased demand for the types of food produced by Australia. However achieving stable government and the resulting economic growth is taking some time, so this trend may not help our terms of trade for some years yet.

Conversely, in the longer term, these countries could become major competitors to Australia's exports. These countries were once known as the bread basket of Europe because of their fertile soils, so once they re-develop their infrastructure and production technologies, they could become large exporters and competitors to Australia again.

Types of change

Serious shifts in markets are generally of two types:

- ❑ medium term change
- ❑ long term change.

Medium term change

A medium term change is a change where there is a shift in the markets because of either over- or undersupply of a product.

The wool industry

A classic example of medium term over supply is currently occurring in the wool industry. Due to many factors, Australia found itself with a huge wool stockpile in the early 1990s. The stockpile was equivalent to a whole season's wool production and it would need a huge increase in demand for that extra supply to be used. This massive over supply meant wool prices fell dramatically and a number of farmers suffered financial hardship. Unfortunately, the over supply coincided with a world wide recession which also limited demand growth in spite of falling prices.

This situation has created a medium term market shift. Wool growers have had to identify that shift and make appropriate changes to their management. Due to this shift, some farmers will suffer business failures.

As mentioned in Chapter 5, there are signs the wool market is picking up again.

The deer industry
In the Australian deer industry, the early growth phase of the industry meant breeding stock were in short supply. This resulted in high prices for breeding stock. As the industry matured and the supply of breeding stock began to match demand, breeder prices fell.

Long term change

Long term shifts are rare but do occur and can cause dramatic financial results on rural businesses.

The citrus industry
An example of this is the Australian citrus industry where for years its traditional outlets were for fresh fruit and juice. In recent times Brazil has become the world's biggest exporter of orange juice concentrate. The Brazilians have developed large areas of rain-fed orange plantings, and with cheap labour they are producing orange juice concentrate at very low prices. This production has resulted in prices being too low for Australian orange juice producers to supply. Hence this market has all but disappeared for local citrus growers.

For local citrus producers to survive they need to move to those markets that provide competitive advantages, such as fresh fruit exports and daily squeezed juice.

The wine industry
The wine industry is another industry that is experiencing significant market changes. Over recent years its export markets particularly to Europe, have grown substantially. For example, exports only took up 3.2 per cent of total production in 1985/86 but by 1991/92 they made up 15.3 per cent of production.

The difficulty is deciding whether this is going to be a medium term or a permanent shift. The only way to answer this question is to assess whether or not this export demand is long term.

Practical aspects of handling risks

There are several aspects of handling risks associated with long term changes in commodity prices.

Being continually alert

One of the major aspects of managing medium to long term risks is to continually assess and monitor developments. Even then it can be difficult to make sound judgements.

It pays to mentally step back from farm operations and view them from a distance while you form a fresh long term view of your business activities.

The key to managing long term change is to continually review the industries in which you are operating.

Increasing the size of the farm

During the 1980s there was a catch-cry of 'get big or get out'. This was meant to be a strategy for counteracting the effects of declining farm terms of trade. Some farmers pursued this strategy with success, became more efficient and stayed viable. Others experienced more risk because their increased borrowings were subject to an unexpected increase in interest rates. At the same time as interest rates rose, commodity prices and seasonal conditions moved against these farmers, causing them even more problems.

Small but exotic industries are susceptible to more sudden mood swings in the market

Pursuing a strategy of increasing farm size obviously requires a great deal of financial planning and timing to be successful, so if you are thinking of increasing the farm's size, give this strategy the attention it rightly deserves.

Increasing farm productivity

This option is generally cheaper to pursue and could provide a reduced risk strategy by improving the farm's profitability in the face of continuing declines in farm terms of trade.

Diversifying

Like short term shifts in markets, it is difficult to predict long term market movements. One of the best strategies to avoid long term market shifts is not to have all your production in the one enterprise. Diversifying your agricultural activities will mean a long term price shift in one market is less likely to affect your overall farm income.

Off farm income

Increasingly more farmers are seeking off farm income. This may include members of the family earning part or full-time income from jobs in the district. This trend has been increasing in Australia and has been a characteristic of US agriculture for some time.

Other forms of off farm income include real estate investments, investments in the share market or investments in local businesses.

7 Managing assets

As was shown in Chapter 2 there are two financial measures that gauge farm financial performance:

- ❑ profit
- ❑ net assets.

The protection and growth of business assets is a very important factor in maintaining farm prosperity and wealth—while avoiding excessive debt is just as important.

The assets of a farm are those parts of a farm that can be sold. They include land and improvements, livestock, machinery, and cash on hand. Table 7.1 shows the assets and their estimated value of the example farm shown in Chapter 2. This farm's assets are valued at approximately $760 000.

These assets have obvious value in providing the resources needed to operate a farm business. However they have an additional value in that they allow farmers to borrow money against them for projects such as buying more land, purchasing farm machinery, getting overdrafts or investing in off farm activities.

The importance of farm assets

If the cash flow and profits of a farm are going well and the bank manager is happy, is there a need to be concerned about the capital base of the farm? It depends on your business objectives.

If your objective for farming is to obtain a profit from the farm that will maintain the family's current lifestyle, then there may not be any need to consider asset values or growth in assets. However if your objective is to maintain a farm that will be able to handle financial shocks over time, such

Table 7.1 *Example assets and liabilities*

Assets	*$*	*Liabilities*	*$*
Land & improvements 245ha @$1235/ha 319ha @ $930/ha	599 245	Bank overdraft Term loan	22 000 55 000
Machinery	130 000		
Livestock sheep	8 000	Other debts	
Stores			
Equity in barley pools	4 200		
Investments Bank shares	20 000		
Total Assets Equity	761 445 684 445	Total Liabilities Equity %	77 000 89.9%

as severe drought or commodity price fall, or having a viable farm to pass onto the next generation, then asset values may be very important.

Therefore, there are three main reasons to be concerned about the farm asset value:

- ❑ maintaining farm viability over time
- ❑ allowing for the financial shocks that occur in agriculture
- ❑ maintaining wealth to pass onto the next generation.

Maintaining viability through periods of economic downturn

The capital you own in your farm, that is the money you would have if the farm were sold today and all your debts were paid off, can be viewed as your safety net against economic hardship.

This is how your bank views your equity when they lend you money.

For example, imagine your farm experienced a drought, which meant you needed $50 000 to cover your operating costs for the next year. If you didn't have this money in reserve, the bank could lend it to you. However they need to know their money is being lent to a secure business. The bank is interested in the farm's ability to repay and if the farm needs to be sold, whether there is enough owner's equity available to cover the added $50 000 debt.

Therefore if there is sound owner's equity in the farm, say 90 per cent equity, the farm is in an excellent position to withstand the short term financial shocks caused by drought and commodity price movements. If the farm's equity is only around 60 per cent, it may still be profitable, but there is not much financial security in the operation. One serious drought or market downturn could send this farm broke. If your farm has low equity, a major objective to cutting risk could be to improve its assets and hence raise its equity.

Maintaining a viable farm to pass on to the next generation

Due mainly to the long term trend of declining farm terms of trade, Australian farms have been under pressure to get bigger. There has therefore been pressure to increase the farm's capital base. With the expected continual declining terms of trade, this pressure is expected to be maintained. If this is the case, then farmers who wish to pass on viable farms to following generations will need to consider how an increased farm size is financially achievable.

The task of increasing the farm's assets is very challenging and requires a lot of careful financial planning. If debt is to be used for land purchase, care is needed to contain borrowing within affordable limits. Tighter pressure on farm profits in recent times has meant farms cannot afford to borrow to the same extent they did 20 years ago.

The continual erosion of farm profits means farmers will often be forced to do things differently. This problem of asset accumulation may require broader thinking. There are farmers who have recognised that financial return for management is more profitable than the financial return from owning land. That is, they have obtained off farm investors to acquire added land and then lease back that extra land from those investors. This has several advantages:

- ❑ it provides a larger land base for the farm's operations
- ❑ it spreads the overhead costs
- ❑ it frees the farm's capital to be invested elsewhere.

Farm assets

Land

Land is usually the most valuable asset of a farm.

As you can see from the farm's asset and liability budget 78.7 per cent of the assets are related to land and improvements. Most people believe that land provides the most stable asset and is the one asset that will tend to best hold its value.

As mentioned, banks regard assets as valuable security to lend against. They particularly favour land. If you borrow money from a financial institution (for example a bank) and secure your lending (mortgage the loan) against your land, you are effectively saying that you are prepared to take the risk that the bank, if necessary, could sell your farm to get back any unpaid borrowed funds. This is a fact that some people seem to forget when they demonstrate against banks who are demanding full repayment from financially troubled farmers.

While banks can be quite reasonable in helping farmers work through their financial difficulties, it is still a major risk to secure lending by putting up land as a guarantee. The real risk occurs when such decisions are made at the discretion of the bank and are beyond the influence of the farmer.

The best strategy against a bank demanding payment is to be totally honest and tell them of any financial difficulties you have. Remember, it is their borrowed funds they are concerned with, and what may be good for you might also be good for them. The last activity banks want to be involved with is repossessing land from farmers. Reposessions are very bad news and very bad public relations for banks. So the best strategy is to try and work with your banker, not against him or her.

Land value

While land is seen as a secure asset, land prices do vary, so it is important to understand why these prices move.

Farmer confidence about the future financial return from rural land is the main cause of land price variation. Farmer confidence can be affected in a number of ways:

- ❑ medium term commodity price movements
- ❑ drought
- ❑ a run of poor seasons
- ❑ prolonged high interest rates
- ❑ government legislation restricting land use.

A classic example of land price variations occurred as a result of the dramatic fall in wool prices. Figure 7.1 shows the major impact wool prices have had on land prices in the South East and Kangaroo Island areas of South Australia. The price movements are shown as an index. These areas are very dependent on income from wool, especially Kangaroo Island, where there are few alternatives at present.

The graph shows the lagged response between wool prices and land values. Wool prices began to strengthen from 1986 through to 1988. Consequently land prices improved from 1987 to 1989. It took from one to two years for farmer confidence to pick up on the strength of the wool price increase.

Conversely it also took some time for farmer confidence to fall when wool prices fell and land prices did not drop for some time.

Movement in land prices can also be influenced by how tightly the land is held within a district. For example, during a prolonged period of poor seasons in South Australia's Upper Eyre Peninsula in the late 1980s, a par-

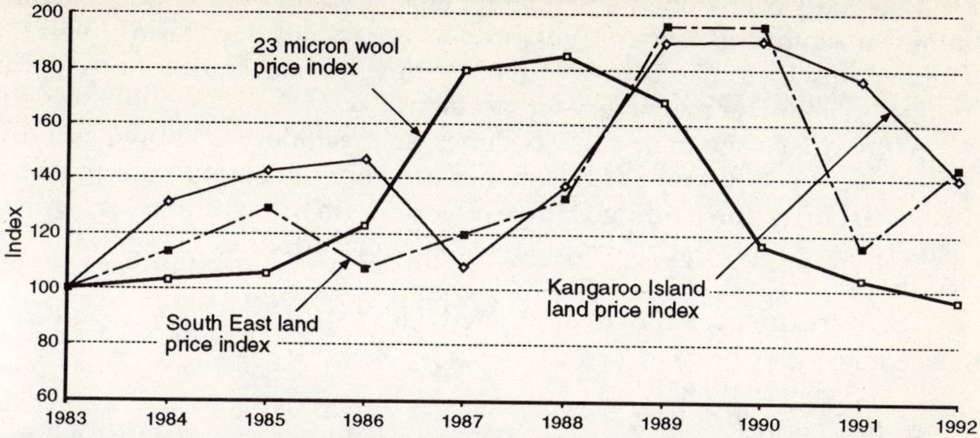

Figure 7.1 *The relationship between land and wool prices in Kangaroo Island and the South East of South Australia*

ticular district called Buckleboo maintained sound land prices even though the rest of Upper Eyre Peninsula experienced land price falls. This happened because an exceptionally high number of families wanted to stay on the land and expand to provide for the next generation.

In other areas where a number of agricultural enterprises can be carried out the impact of commodity price movements is not as direct.

Figure 7.2 shows movements in the land price index for Yorke Peninsula and the Lower North areas of South Australia for the twelve years 1980–92. Over this period land prices have strengthened, although farmer confidence has been affected by movements in interest rates and wool prices.

Machinery

Depending on the type of farm, machinery can be the next biggest business asset. This is the case for most mixed farms where a typical farm business could have about 17 per cent of its assets tied up in machinery.

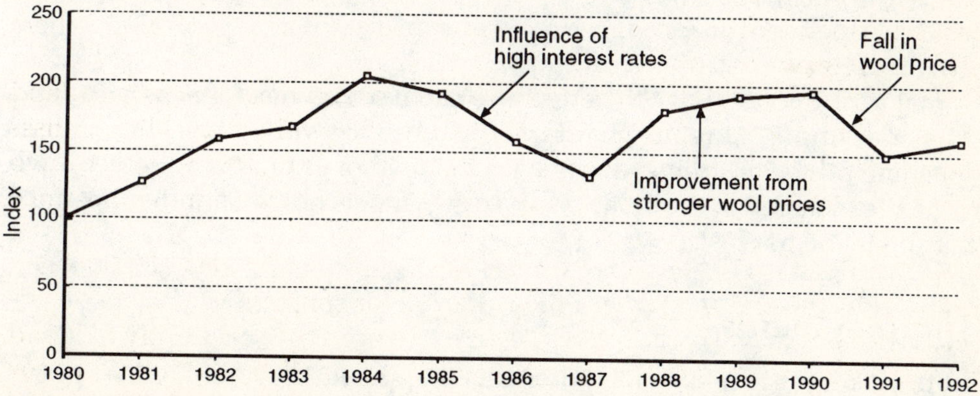

Figure 7.2 *Land price index for Yorke Peninsula and the Lower North of South Australia*

Machinery is obviously essential for efficient farm production, so maintaining machinery is therefore very important for continued farm efficiency. Unlike land, there needs to be a replenishment factor built into the farm budget to allow for machinery replacement.

As machines age their value declines, so machinery is known as a depreciating asset. Because of this, banks are very reluctant to use machinery as security for lending. However, some finance companies, which usually have more expensive finance, may consider machinery value as lending security.

There are three main risks associated with machinery:

- ❑ machinery break-down
- ❑ machinery theft
- ❑ fire damage.

Machinery break-down

The timeliness of machinery operations, particularly at seeding, spraying and harvesting periods can be very influential on productivity and hence farm profits. Some examples include:

- ❑ a delay in seeding by three weeks, due to a major tractor breakdown, can reduce yields significantly
- ❑ cattle musters can be interrupted if helicopters are out of action for weeks awaiting spare parts.

There is clearly a risk that some farm operations may be interrupted by machinery breakdowns, so carefully maintaining machinery and regularly servicing it can help cut risks.

Machinery theft

Machinery theft, while less common in rural areas compared with car theft in urban areas, can create two types of risks:

- ❑ the risk of capital loss
- ❑ the possible cost of not being able to perform the farm operation at the optimum time.

Fire damage

A major risk occurs with the effect fire can have on machinery.

For example, a fire may be started in a header while harvesting and the machine can be destroyed along with the field of grain. An added cost also could be incurred in the delay while a farmer gets hold of another header to complete the harvest.

Improvements

Improvements to a farm include items like sheds, the house, water points, fences, contour banks and stock shelters. Major risks to improvements are

Machinery break-down will cause delays and extra costs

from flood, fire and asset deterioration through age and use. Termites and mice also can cause improvement deterioration and crop damage.

Livestock

Livestock on most mixed farms forms a relatively small part of the asset base and usually only stock and station agents will lend carry on finance against the value of stock—except in the case of grazing properties where stock form the biggest asset base after land.

The major risks associated with livestock ownership are:

- fire
- drought
- theft
- pest and disease outbreaks
- price fluctuations
- birth losses—especially in stud stock.

The enterprise mix on your property will decide how important livestock are to your farm profits. The risks associated to the livestock enterprise will also depend on the type of livestock being carried. For example, drought would have little effect on the health of housed poultry, but a disease outbreak could be disastrous. Likewise, drought could be devastating for a cattle or sheep station, but a disease outbreak may only affect a small portion of the stock.

The risks listed above can be disastrous not only for the capital value of the business, but also for lost profits. Capital losses can be very substantial if stud stock are lost. The extent of the risks and their likely impact on the property are very dependent on the type of livestock, the location of the property and the management of the property.

Manager's life and disability

The manager of each property is a key asset. Although not listed in the balance sheet, he or she has a vital role to play in the success of the operation.

What would happen to the business if the manager were permanently or temporarily disabled, or died? Would the profits of the business suffer?

The strategy to manage this risk depends on the objectives of the owners of the business. That is, whether the property is to be sold or retained in the event of death or disability.

Sometimes banks insist life and disability insurance be taken out by the manager before they will lend. This insurance is to cover the bank's debt in case the manager has an accident which results in disability or death. Farmers can take out similar insurance, as their dependants may need to employ a manager in the event of the manager's untimely death or disability.

Practical aspects of handling asset risks

Land value

There are a few practical management techniques that can guard against land value movements.

Properties in different districts

Having land holdings in different districts may spread risks.

It is likely different areas would experience different seasons and could also operate different forms of rural enterprise. However you will need to balance this strategy with the logistics of managing two properties that may be some distance apart.

An historic example of this is the Kidman cattle empire where cattle stations where spread widely across the Australian outback. This provided not only a practical drought strategy but also spread the owner's asset base.

Sound relationships with your bank manager

If you have a mortgage against your land, the best way to minimise the

Accurate farm records and a well developed plan will aid your case when it comes to negotiating a bank loan

bank requiring quick debt repayment is to maintain an excellent working relationship with your bank manager. This can be done by:

- being completely honest in all your financial dealings with the bank
- remaining in regular contact to keep the bank informed of your current and likely financial position
- maintaining effective financial management, which concentrates on forward projections and planning.

Keep a high level of equity in your business

If high equity is maintained on the property (90 per cent equity or above) then there is leeway for both commodity price movements and adverse seasonal influences. Your bank will also feel safer lending to you. A high equity will also allow you to be flexible with your farm activities and future land acquisition plans.

Have some off farm income

Not having all your eggs in one basket and having off farm income, will mean your risks have been spread beyond the farm and you are unlikely be adversely affected by any drop in land prices.

Is education an asset?

Does being a trained teacher, a nurse or a qualified trades-person provide an asset to the farm business? For farm families looking for off farm income, education could be considered an intangible asset (an asset that can not be sold but only hired). However as education provides a key to earning, it is an asset that can provide stability to the business income and reduce reliance on other income earning assets.

Machinery strategies

There are several strategies that can be adopted to minimise risks associated with machinery.

Machinery breakdown

The best strategy against mechanical breakdown is to either maintain a sound repair and maintenance program, or regularly replace machinery. You can also purchase only those machines that are served by competent local service agents and who provide a good spare parts service. You may also decide to carry essential spare parts on the farm. For example if all your tractors are brand X then developing your own repair skills and keeping a stock of spares is easier than if you have several different brands of X, Y and Z. This is especially important if you are a long way from service centres like some remote properties in western and northern Queensland, northern Western Australia, northern South Australia and the Northern Territory.

Machinery theft

Machinery theft can be prevented by taking keys out of vehicles and maintaining a lock-up site for especially valuable machinery. Perhaps the best strategy is to insure against theft. Some districts have established a Farmwatch service where neighbours can report strange vehicles either to each other or the police.

Fire destroying machinery

Preventative maintenance will reduce the risk of machinery causing fires. However there is always fire risk through Australia's summer months due to heat, dry conditions and thunderstorms.

The most common strategy to cover the risk of property loss by fire is to have a fire insurance policy. You should also carry fire extinguishers on machinery and check them regularly to ensure they are working properly.

Improvements

The major and most common strategy to cover financial loss of farm improvements is through normal farm insurance. Such insurance policies can cover theft, fire and flood.

It pays to shop around as premiums can vary a lot. Make sure too that you carefully read the fine print so you know exactly what risks have been covered.

As fire is the main risk to property improvements, a fire prevention strategy will help you cut the risk of fire. This strategy might include making firebreaks, getting fire fighting appliances, obtaining good communications equipment, and siting irrigated paddocks near buildings.

Livestock strategies

There are several strategies that can help cut risks to livestock.

Fire
Managing fire outbreaks is as much a community issue as a farm issue. Precautions such as having adequate fire units on hand, maintaining fire breaks and having sound communications networks are all parts of a sound fire fighting strategy.

Farmers can take out insurance to cover the value of livestock in the event fire causes stock losses. Moving livestock onto lucerne and turning on irrigation may save them from fires. This action may be a worthwhile precaution for valuable stud stock and is a humane action for all animals.

Drought
Drought is a real risk for livestock, especially in the rangeland areas of Australia. There are strategies designed to cut the risk of drought which have already been covered in Chapter 3 of this book.

Theft
Theft of livestock is slightly different from theft of machinery, as it is more difficult to police and securely lock up livestock. However precautionary measures can be taken such as taking part in the police and community operations such as Farmwatch.

Again, theft insurance can be taken out to cover the cost of livestock losses. Locked gates on roadsides may deter thieves, as Farmwatch participants are likely to notice and report strangers cutting fences and so on.

Pests and diseases
The type of livestock and severity of the disease, will determine what strategies need to be adopted. There are government quarantine and treatment guidelines laid down in case of exotic disease outbreaks. In some instances whole herds and flocks may have to be destroyed.

Table 7.2 outlines the types of exotic diseases where governments will provide a form of producer compensation. This compensation covers you for the replacement of animals that have to be destroyed due to an exotic disease outbreak. In other countries there are insurance policies farmers can take out to compensate for loss due to exotic diseases. This type of insurance is not available in Australia.

In the case of minor diseases there are veterinarians who can treat animals. Also there are precautionary measures livestock owners can take to

Table 7.2 *Exotic diseases covered by government compensation*

Foot and Mouth Disease	Rinderpest
Vesicular Stomatitis	Bluetongue
Swine Vesicular Disease	Avian Influenza (Fowl Plague)
Vesicular Exanthema	Newcastle Disease
Classical Swine Fever (Ho Cholera)	Rabies
African Swine Fever	Screw Worm Fly

help prevent pest and disease outbreaks. Such measures include vaccinating and drenching stock.

Livestock prices

Livestock prices can be volatile and will be influenced by both the domestic and export markets, depending on which market is being used. Short term price stability may be managed by forward physical selling, or if available beef and wool futures markets can be used. However for most livestock there is no mechanism to help farmers stabilise either short or long term prices.

Managing life and disability

There are many strategies available to deal with the risk of a manager's disability or death, but the right strategy depends on the objectives of the business.

If the objective is to maintain the property, then enough money needs to be either placed aside or insured for, to enable you to hire a replacement manager for as long as necessary.

The property may also be operated on a lease or share farm arrangement, which would enable the business to continue. However profits may be lower, depending on the performance of the new manager.

The other alternative is to sell the property and invest the proceeds to help support the family.

Many women on farms are studying farm management and working with their partners, so they may well be able to manage the property if left alone.

8 Financial risks

The major risk in borrowing money comes from giving the bank (or lending institution) the right to decide your business destiny if you cannot meet your repayments. If the bank feels its investment in your business (the money lent to you) is threatened, they may act to protect their depositors' money before they consider you.

The other related risk is fluctuating interest rates.

Fluctuating interest rates can make profits and viability uncertain. However, with the freeing of the finance industry, a number of financial tools have become available to help manage these interest rate changes.

Correctly using borrowed funds is a good way to secure wealth and capital for a farm business. However, effectively managing these funds is the key to using such funds successfully and to make sure they benefit your business to the maximum. One of the best ways to learn how to effectively handle borrowed funds is to learn from those successful farmers in your district who have borrowed funds before.

The Australian Bankers Association has produced a guide called *Financing your farm—a practical guide to financial planning and management.* This guide also outlines how to manage borrowed money.

The interest rate environment

Since the finance industry was deregulated in the early 1980s, we have seen interest rates vary widely. This deregulation brought with it more lending flexibility and borrowing risks. These risks were highlighted by the large numbers of farm bankruptcies that occurred through the early 1990s. These increased risks have meant managing borrowed funds has become very important.

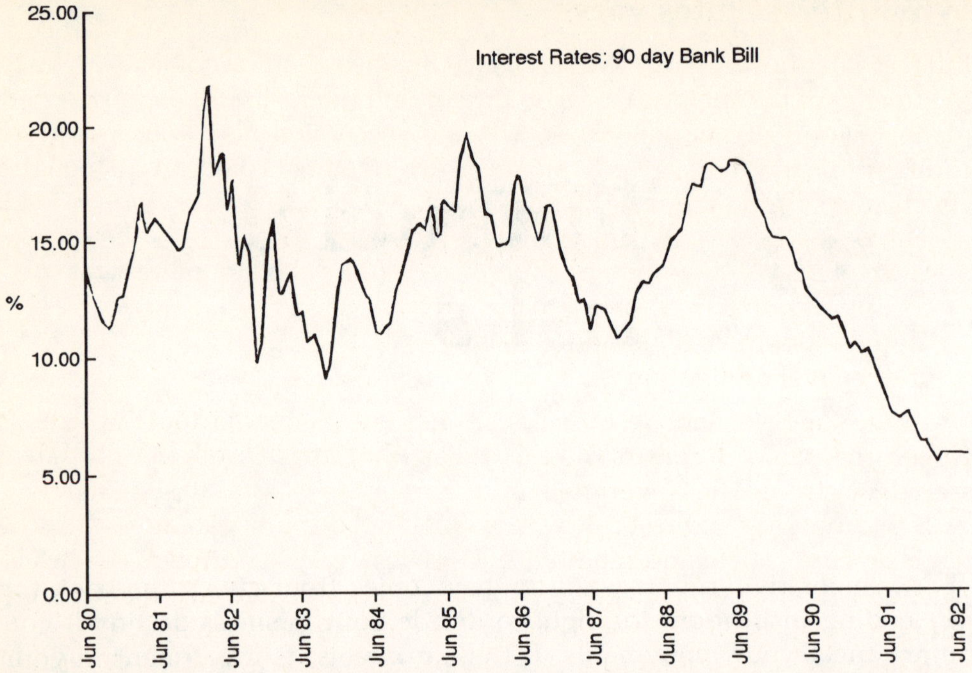

Figure 8.1 *Short term interest rates.*

Lending can now be made through fixed or variable interest rate loans. Managing variable interest rates helps cut the risk of borrowing money. The general lending products available from finance institutions are shown.

Using a combination of these funding arrangements helps spread the risk of unpleasant interest rate movements.

The topic of finance can be quite complex and one of the most difficult aspects of finance is to understand the terminology used. Stick with learning these terms, because you will find it very useful in your future business dealings. Seek out a friendly bank manager to help familiarise you with managing finance or do a course on rural finance.

Table 8.1 *Types of loan*

Type of Loan	*Interest rate*	*Repayments*
overdraft	variable	interest and principal
term loans	variable	interest and principle
bank bills	fixed	interest only
fixed term fixed interest	fixed	interest only
maximum and minimum interest rates (caps and collars)	a combination of fixed and variable	interest only
concessional lending through the Rural Finance Development Division (name varies in each state)	fixed initially then reviewed annually	interest and principal

Why interest rates vary

Interest rate movements have hurt farm businesses in recent years. Due to the freeing of the financial market, fluctuating interest rates are now a permanent feature of our economy. However deregulation has encouraged more competition and there are now financial arrangements that can cushion the effects of changing interest rates.

The major factors that affect interest rates are:

- ❑ the Federal Government's monetary policy (setting interest rates)
- ❑ international interest rates
- ❑ the economy's demand for money
- ❑ the level of inflation.

As we suggested above, monetary policy is an economic tool the Federal Government uses to control the Australian economy. If economic growth is seen to be strong, the Government may raise interest rates to slow it down.

If international interest rates are rising, then our interest rates will have to go up to compete and maintain the level of overseas money staying in Australia. If our rates stayed low then investors would take their money out of Australia and move it where interest rates were higher.

If the economy's demand for borrowed finance increases, borrowers will compete for those funds and so pay higher interest rates to get those funds.

Inflation also has an indirect effect on interest rates, as depositors want higher interest rates during times of high inflation. This allows them to maintain the real value of their investments.

Figure 8.2 shows how both long and short term borrowing have moved over the last twelve years. At some stages, long term rates are higher than short term rates. The opposite also occurs. Based on the observed interest rate cycles, farmers can form strategies that will help cut the risks of high interest rates.

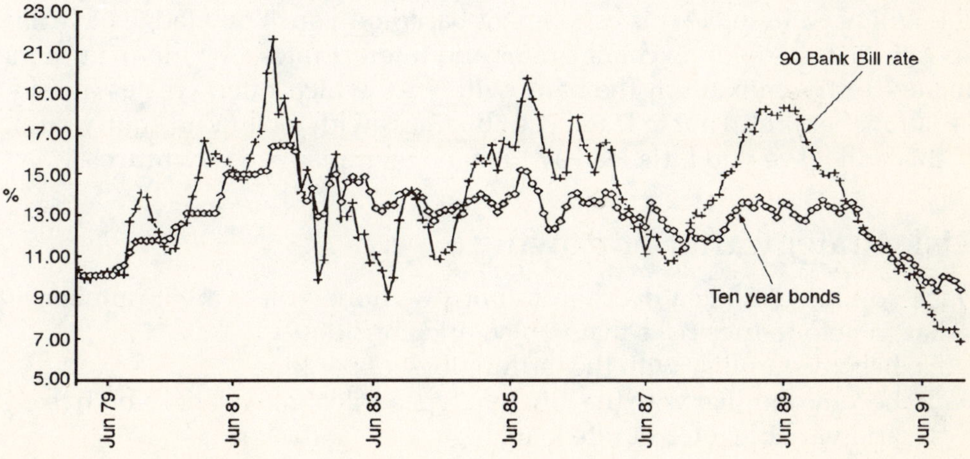

Figure 8.2 *Australian short and long term interest rates.*

Practical ways of handling risks

There are several practical ways of handling risks and we mention some of these below.

General strategies

New financial arrangements can include:

- ❑ fixed interest rate periods
- ❑ interest-only loans
- ❑ any combination of these.

When borrowing it is important to consider:

- ❑ your business' ability to repay
- ❑ how you expect interest rates to move
- ❑ any taxation implications.

You should take these factors into account before deciding on the right method of financing your farm business. Also, you should always consult your bank manager and accountant before making any big financial decisions.

The major part of any strategy is to be aware of the profit strengths of your own farm business and the financial arrangements that are most suited to it. If your business cannot afford to speculate with fluctuating interest rates seek advice from your banker and your accountant, because fixed interest rate loans are an option to give some protection against interest rate fluctuations.

Risk strategy before borrowing

Understand your property's financial strengths and weaknesses. Make sure you know how much borrowed money your business can safely support. You will need to make this assessment based on your knowledge of variations in prices, yields, exchange rates and interest rates. Too many farmers borrow to the maximum the bank will lend, which often creates a risky situation. Be careful not to leave your business with too low an equity ratio, as this will leave you little leeway to manage unplanned downturns.

Risk strategy after borrowing

Once you have made a decision to borrow money there are a number of issues to help reduce risks that you should consider:

- ❑ become familiar with the terminology of banking
- ❑ become familiar with the different types of loans available—both fixed and variable interest rate loans
- ❑ use a manager and a lending institution that understand your busi-

ness and have confidence in you. If you foster a good relationship with such a person they will more likely stick by you through any tough times.

- ❑ have a strategy of mixing your borrowing between fixed and variable interest rates
- ❑ monitor the interest rate outlook.

Cutting borrowing risks

When borrowing money, it is best to gain the advice of a sound financial adviser and get a spread of both variable and fixed lending.

As the 90 day bank bill (Figure 8.1) indicates, the interest rate cycle since deregulation has been about three years, so if you are locking into an interest rate, it is advisable not to lock into periods of more than about three years. See your financial adviser on this issue when you are planning to borrow more funds.

Although interest rates tend to follow cycles, the future is always uncertain. It is wise to spread your risks between a mixture of fluctuating and fixed interest rates, rather than try and pick a winner and attempt to guess what rates will be. In the long term the interest rates you pay using this method should be lower than if you don't spread your risks. The following example illustrates how this strategy works.

In this example farm the owners are going to borrow $110 000 to buy land and machinery and $40 000 for carry-on purposes—giving a grand total of $150 000. The spread of the borrowing strategies are shown in Table 8.2. With a general interest rate of 14 per cent, this farm's annual interest repayments will be $18 200.

If the interest rates increased to 18 per cent for the year, the farm's interest payments would have increased to $21 000, as shown in Table 8.3—assuming the bank bill interest rates remain fixed at 14 per cent.

The rise in interest rates caused a jump of $2800 in interest payments. If all borrowing was made with variable interest rates, the annual interest bill would have increased to $23 400—an increase of $2400 over the spread borrowing strategy.

Table 8.2 *Lending example at 14 per cent*

Interest rate	*Borrowings*	*Interest Payments*
14 %	Bank bills $60 000 @ 14% (interest rate fixed at 14 %)	$8 400
	Term loan $50 000 @ 14%	$7 000
	Overdraft $ 40 000 @ 14 % (for 6 months only)	$2 800
	Total	$18 200

Table 8.3 *Lending example at 18 per cent*

Interest rate	*Borrowings*	*Interest Payments*
interest rates move up to 18%	Bank bills $60 000 @ 14% (interest rate fixed at 14 %)	$8 400
	Term loan $50 000 @ 18%	$9 000
	Overdraft $ 40 000 @ 18 % (for 6 months only)	$3 600
	Total	$21 000

If interest rates were to fall to an average of 10 per cent, Table 8.4 shows the financial results for this borrowing strategy—assuming the bank bill interest rates still remain fixed at 14 per cent.

A falling interest rate meant the interest payment dropped from $18 200 to $15 400, a saving of $2800.

If no spreading of interest rate risk was done, the savings would have actually been $5200. This shows there is a risk in locking in interest rates in a falling interest rate market.

However the major advantage with spreading the borrowing between fixed and variable rates is that it helps reduce the interest rate burden associated with interest rate increases. This is very important to rural businesses that cannot afford such increases.

If our example farm was eligible for concessional lending then the debt burden would be further cut with the lower interest rates, using government assistance lending.

Table 8.4 *Lending example at 10 per cent*

Interest rate	*Borrowings*	*Interest Payments*
interest rates move down to 10%	Bank bills $60 000 @ 14% (interest rate fixed at 14 %)	$8 400
	Term loan $50 000 @ 10%	$5 000
	Overdraft $ 40 000 @ 10 % (for 6 months only)	$2 000
	Total	$15 400

Consider the long term implications of borrowing money, especially shifts in interest rates

9 Biological risks

Biological risks are common risks that are managed by farmers with varying degrees of success. With the continual release of chemical sprays and improved technology, some biological risks are becoming more and more manageable, but others still remain problems.

While this chapter provides a brief discussion on how to manage these risks, there are many more practical management strategies that you can explore. Technical information on these strategies can be obtained from your local Department of Agriculture or Primary Industries.

Biological risks include those biological events that cause uncertainty. They may include the effects of vertebrate and invertebrate pests, plant and animal diseases, plus weeds. These include:

- vertebrate pests—such as rabbits, mice, dingoes, wild horses, foxes, goats, donkeys and feral cats
- invertebrate pests—such as locusts, sheep blow flies, snails and aphids
- plant diseases—such as cereal rusts, chocolate spot in beans, cereal cyst nematode, hay-die, tobacco mosaic virus in tomatoes and curly leaf in stone fruit
- animal diseases—such as pulpy kidney, foot rot, botulism, leptospirosis and leucosis
- weeds and other plant pests
- pests of animals—such as ticks, lice, worms and hydatids.

Some farmers may argue biological risk should also include risks such as salt affected land caused by over clearing. We cover land degradation in Chapter 10.

Examples of biological risks

Risk brought on by biological factors can be quite unpredictable and a few examples are listed below.

Cereal root diseases
These diseases cause a great deal of yield loss in cereals by restricting the water and nutrient uptake of the plant.

Downy and powdery mildews
Downy and powdery mildew are fungi that infect many fruit crops, affecting leaves and fruit. Mildews cause problems, especially in grapes.

Stem rust
This type of rust affects the stems of cereal plants by restricting the flow of moisture and nutrients into developing grains. This affects yields.

Locusts
Locusts have the potential to eat any green material in their path. Hence, if not checked early in their life cycles, they can completely eat crops and pastures over large areas. They also can fly large distances in a short time.

Fruit fly
Fruit flies are serious threats to special export markets in those States that are free of the pest.

Salvation Jane / Patterson's Curse
This weed, *Echium plantagineum*, was introduced as a garden flower from southern Europe last century. It took to southern Australia so well it has become a widespread weed on many farms. In its early stages it is palatable to stock but when mature *Echium plantagineum* can be toxic. This creates risks for the grazing industry.

Barley grass
Barley grass is a weed of some southern Australian cereal country. It is viewed by some farmers as providing good grazing in its early growth stages. Once it matures its grazing value is poor. Also it competes vigorously with cereal crops. Being a grass, it has the potential to host and multiply cereal root diseases, which may affect yields of following cereal crops.

Lantana
Lantana is an invasive and difficult-to-control weed of Australia's sub-tropics. Like *Echium plantagineum*, it was introduced as a garden plant from overseas.

Prickly acacia

Like lantana, prickly acacia was introduced from its native range in central and eastern Africa into the Charters Towers district of Queensland last century. Now stands of dense prickly acacia are spreading across the inland from Charters Towers west to Mt Isa.

Rubber vine

Rubber vine was also introduced into Australia from overseas. The weed has spread over large parts of north Queensland and the Northern Territory where it seriously threatens native forests and rangelands.

Starlings

Starlings occur throughout Australia except in Western Australia. These introduced pests damage fruit crops and other valuable products. The West Australian government wages an active campaign to prevent the birds crossing the Nullarbor from neighbouring South Australia.

Rabbits

The introduction of rabbits by Europeans into Australia has presented one of the largest risks to the land. Rabbits can increase to plague proportions. They not only compete directly with livestock for grazing, but have displaced bilbies and other native marsupials from their home ranges. Rabbits can degrade land by baring soil, so allowing water and wind to erode the land.

Some introduced species can cause great economic damage if they reach plague numbers

Exotic disease outbreaks
Exotic disease outbreaks are feared by both individual farmers and rural industries alike.

When such outbreaks occur, farmers generally have all their stock or land quarantined to help contain the outbreak. Once contained, the exotic disease must be eradicated, usually by destroying affected or susceptible stock or plants. Individual farmers can lose large amounts of money which may threaten the property's viability.

The major exotic disease risk to an industry is through loss of markets. Both domestic and international markets, may be closed to all Australian farmers until Australia can show it has eradicated the disease.

The are many other examples of biological risks. Farm managers should be aware of the importance of all biological risks and know which pose the greatest financial threat to their farm businesses. These risks should then receive the greatest attention.

Practical aspects of handling biological risks

There are a number of management strategies that can be used to help cushion farm businesses from some biological risks.

Integrated pest management (IPM)

The incidence and impact of some pests, weeds and diseases can be managed with a mixture of chemicals, predators and parasites. These options are numerous and a lot of technical advice is available from both governmental and commercial sources.

Be aware of the costs of these treatments and the broader problem of chemical resistance that can be encouraged if the same type or group of chemical control is used consistently.

In some industries like cotton and lucerne-seed production, specialist pest control consultants help in pest management. These services help yield loss risks.

Equity levels and cash reserves

One of the main strategies for withstanding the financial effects of yield loss through pest and disease problems is to have a high equity in the property. This strategy is very useful for all risks confronting farmers and hence is the most common strategy mentioned throughout this book.

As previously mentioned, a third of Australia's farms today are debt free, so obviously they have this biological risk strategy available to them.

Insurance

Insurance policies in general provide cover only for weather related risks, such as hail, frost and wind damage. While some diseases and pests are encouraged by certain weather conditions, generally it is not possible to take out insurance for these.

One exception is bunch rot and botrytis in grape vines. These diseases depend very much on weather conditions through the summer months. Farmers can insure to cover some of their expenses if these diseases should occur.

Another exception is for endemic disease outbreaks in cattle feedlots. The feedlot must have strict, high quality hygiene and management standards before an insurance company will consider issuing a policy.

Exotic disease outbreaks

Currently, if an exotic disease outbreak occurs, the property is quarantined and the livestock or crops are destroyed. Either State or Federal Governments will provide compensation to the property owner for the destroyed stock or crops.

There is very little a property owner can do, either to prevent an exotic disease outbreak or minimise the damage once it has occurred. Fortunately there have been only a handful of exotic disease outbreaks in Australia, so the chances of your property being exposed is remote. The Australian Quarantine Inspection Service (AQIS) handles these risks for farmers and society in general.

New technologies

There are new technologies being developed continually to help prevent biological risk impacting on rural properties. One that is being adopted by farmers is break crop technology. This helps control cereal cyst nematodes in cereals. Here a two year grass-free crop or pasture is grown between cereal years.

10 Preserving your land

Most Australian farmers are probably aware of the Landcare program and how it is changing attitudes of Australian people to caring for and managing farm land. Land care is an important issue for the rural people, but farm managers must make sure the projects they undertake are within the financial capacity and risk profile of the property they run. These decisions must therefore be taken with a great deal of planning and careful analysis.

Land care planning and financial analysis can be quite involved. Therefore getting some training or using land care consultants will help you develop a sound plan. This approach also will help you monitor and manage the risks sometimes associated with individual property land care projects.

The Landcare program

The Landcare program arose from a need to focus people's attention onto one of Australia's greatest natural resources—its land.

Australian farm land has been scoured by wind and water erosion since first settlement. Our land has also been degraded by rising water tables bringing salt into root zones. Soil fertility is falling on some properties due to erosion and continuous heavy cropping. These and other problems have affected many farms.

Caring for the land is now an important part of farming and needs to be part of any sound farm management program. The challenge is to sort out:

- ❑ the priorities of potential land care projects on each farm
- ❑ the practical and financial risks of projects
- ❑ a strategy that will maintain financial viability while improving land quality.

This program should involve identifying all the likely risks and deciding on those land care projects which would be most effective to minimise those risks. The Landcare program poses its own set of challenges to property managers.

Financial viability versus land care

The dilemma for most farmers is to improve their land care management while still remaining financially viable. Most land care activities on farms incur increased costs and could decrease productivity for some time. Such activities include:

- ❑ planting shelter belts and wind breaks
- ❑ developing contour banks
- ❑ refencing to better control stock in pastoral country
- ❑ refencing to better suit soil types in cropping country.

All these projects need money and all would disrupt production in their early years. These activities might therefore cut incomes in their initial stages.

All these land care risk aspects need to be considered, as they cause increased risks to farm businesses. In some instances there may be a difficult trade-off. For example, a land care farm project may provide short term risks but if the project is not carried out it might create a bigger risk in the future. An example is a gradual decline in productivity from salt affected areas. Here, rising water tables need to be controlled, otherwise there may be continual land deterioration.

Not all farm land care projects conflict with farm viability. Some will actually increase productivity and ensure long term viability. An example might be improving soil management by reorganising fences to better match soil types.

Improved cropping rotations can also help improve soil fertility, improve soil structure and increase productivity.

Land care failure

Some management strategies used to solve certain land care problems are not proven. This is another risk farmers face.

For example, one management strategy to help prevent water tables rising is to re-establish trees on affected properties.

Because the rising water table problem is a relatively new land care issue, and tree planting is a long term solution, there is a risk this strategy may either not work or be too late to save some land.

The money invested in tree planting could be spent with no medium to long term financial reward. This is the risk of using an unproven strategy to overcome a land care problem.

Similarly there are risks associated with the co-operation needed from groups of farms to solve problems of rising water tables in a district. For

The hoof action of some species can contribute to land degradation

example, re-establishing trees to control rising water tables may need all the district's farmers to take part in a tree planting program. If some people do not take part the strategy may fail.

As we know, some farmers will be financially better able to take part in such a tree planting program than others, but the benefits will be felt by many farmers. In this situation, the risk of non co-operation from farmers could have serious consequences for the whole district.

The farmer who didn't change

An interesting true story regarding two farmers highlights the risk and financial difficulty which can be characteristic of land care projects.

An innovative farmer, John, was conducting a rotation and tillage trial on his cereal property. His trial found an improved farming system which he adopted. This new system incorporated reduced tillage techniques, more cropping, a cut in stock numbers and improved pastures. The advantages John received were improved soil structure, better soil fertility, reduced root diseases, improved yields, increased livestock productivity and above all increased profits of 25 per cent. The risk was that it took three years to adopt the improved system and there were reduced profits and increased running costs during this period.

John had a neighbour Bill, with identical soil types. Through field days Bill could have learned all he needed about the new farming system. In addition Bill had the benefit of being able to look over John's fence and see the improvements for himself. For his own reasons Bill didn't change his farming system.

About eight years went by. John's business continued to perform well and Bill still didn't change.

Last year John bought Bill out, as the profits of the old system were not enough to maintain the neighbour's property.

Bill chose to minimise short term risks. His long term risk of declining farm terms of trade increased, which eventually forced Bill to sell out.

Practical aspects of handling land care risks

Land care training programs

Courses are now available on rural property planning which outline the best way to adopt and improve land care programs. They teach farm managers property mapping skills to allow managers to relate soil types to current fence lines. Also they help farmers decide on improved property layouts. Such courses also help in comparing alternatives and selecting the best options for each farm.

Financial management

The best option to minimise land care risks are to fully appreciate the financial position of your property. From that base you can work out where the greatest risks are to your farm's viability. Once you have assessed these risks, it will be easier to prioritise the potential land care projects and fit them in with normal farm activities.

11 The risk of inadequate information

With a continual supply of new research results and improved communications technology, information is becoming more easily obtainable. Some are calling the 1990s the information boom era. Compared with only five years ago, computer technology is cheaper and more user-friendly, fax machines and two way radios are common on farms and mobile phones are also appearing.

The amount of information available means farmers can be swamped.

Often when we are looking for a specific piece of information, such as what chemical to use to control a particular weed, there is so much information that sometimes we can't find what we want among the mass of available data.

We can also experience difficulties when making decisions where the information we need is difficult to find.

For example, research officers may have solved the problem, but may not have made the information readily available to farmers. In these instances, farmers may have to develop innovative ways to gain this information, such as personally contacting research officers for the information.

There is a skill to knowing how to efficiently obtain the necessary information and when you come across conflicting information, knowing what source to trust. The challenge to managing information risk is to continue to operate your business efficiently, while being continually aware of new ideas and technologies. Also it is important to try to understand emerging domestic and international trends. Understanding these trends will require you to continually prioritise the flow of information and to be aware of the risks that come from using or not using that information.

Information risk

Some information is vital to the financial performance of properties. It is the manager's responsibility to decide which are the vital pieces of information. The following examples show where inadequate information increased risks to farmers.

Minimising tax

An example of information risk occurred in the early 1980s when due to a run of good seasons farm returns were sound. In an effort to minimise tax and with the advice of some accountants, a few farmers leased new machinery. The lease payments allowed the purchase of the new machinery to be written off against tax, so tax bills were minimised.

Unfortunately following seasons produced less than average production. We saw poor commodity prices and substantial increases in interest rates. These changes caused farm financial circumstances to change. Incomes fell while lease payments increased and soon those farmers had no income to write off against tax. So what began as a successful short term tax minimisation activity, turned out to create viability risks.

The difficulty was that the information may have been correct, but it did not tell the whole story. Often the most important management skills are to know what questions to ask experts, rather than settling for general answers.

Overseas loans

A further example of information risk was farmers' borrowing from overseas banks in the mid 1980s. When the finance market was deregulated, some banks allowed farmers to fund their debts by borrowing overseas. At that time these borrowing practices provided a new opportunity and seemed to be an exciting way to raise funds and reduce repayments.

The attraction for overseas borrowing was that the interest rates were about a third of Australian rates, so the cost of borrowing funds was greatly reduced. However the catch was that if the $A exchange rate fell, the capital amount initially borrowed would increase and in some circumstances this increase was substantial. To cover this risk the borrower could obtain currency insurance cover. However the irony was that if the exchange rate cover was taken out, the cost of the overseas loan would make it virtually the same as borrowing in Australia!

For a number of reasons a few farmers took out this currency cover and as history shows, the $A exchange rate did drop. As a result, these farmers debts grew enormously, which caused a lot of them to go out of business. Clearly their information on borrowing overseas was deficient and provided added risks for borrowers.

Both professional advisers and farmers had the best intentions when they attempted to minimise tax or cut the cost of borrowing. However these examples show there are risks associated with getting poor or inadequate information. Both professionals and farmers alike cannot foresee the future and mistakes are so easy to make.

Farm consultants can provide advice on the likely profit and risk impact of changes

Information sources

It is important to assess where and how you get the information you need to run your property. With the information boom, a lot of time can be wasted trying to find answers. Efficient information sourcing will become an increasingly important issue and one which managers will need to develop techniques to handle.

Information can be divided into:

- ❑ external information—where information is gathered from outside the farm
- ❑ internal information—where information is obtained from inside the farm.

External information sources

External information sources provide the greatest volume of information and this volume is expected to go on increasing.

A recent study commissioned by Westpac found the major sources of information used by successful farmers were:

- ❑ rural newspapers
- ❑ other farmers
- ❑ Departments of Agriculture

- ❑ market reports
- ❑ field days
- ❑ rural radio.

Most sources of information can be useful, but it takes time to find out the information most pertinent to a property's problems.

Internal information sources

Internal information is that information generated from farm records. It includes records such as:

- ❑ paddock books
- ❑ cashbooks that have recorded what has occurred
- ❑ planning budgets.

This is the area that will require more development in the future, as increased pressure will be placed on farms to be run as efficient businesses. Farmers will need to record activities and decide which activities are most profitable.

Deciding on necessary information

The key to managing all risks is to have the farm's goals and objectives carefully thought out and recorded. This will help focus your managerial effort to achieve those goals. It will also assist you to know:

- ❑ what information you must have
- ❑ what information you should have
- ❑ what information you could have.

It is also wise to know what information carries the most risk.

The risk associated with information cannot be fully eliminated. However by carefully collecting information from different sources, it is possible to cut risks. For example, a weather forecast could be obtained from both public and private sources and compared. Weather records also could be used to check the probability of favourable rain. Obviously a number of sources should be accessed to provide valuable information for decision making.

Identifying important information

Identifying important information is one of the most important factors in managing information. The following example illustrates the need for information. A farm manager needs to decide whether or not to put nitrogen on a crop. The additional nitrogen application will depend on the per tonne price of the crop, the resulting extra quality and the cost of putting on the extra nitrogen. We can get information on all these factors, but which is the right information we need to make the best decision?

Neighbours and local experts are usually good sources of information pertinent to your region

We know:

- ❑ the cost of the extra nitrogen in $/ha
- ❑ the approximate price of the crop delivered to the mill for the coming season
- ❑ the yield response to extra nitrogen
- ❑ we can estimate the rainfall, based on previous records—this rainfall is very uncertain.

Local agronomists advise that the yield response could be anywhere between a loss of 10 per cent if the season was poor, to an increase of 30 per cent with an exceptional finish.

What are the key information areas we need to concentrate on to help make the correct decision? In order to answer that we need to know what profit margin the crop is generating before nitrogen is applied, and how much the yield must increase to pay for the nitrogen. Table 11.1 shows existing income and costs of producing the wheat.

If no nitrogen is applied, the gross margin in Table 11.1 is expected to be $267.20/ha. The cost of the nitrogen application is $15/ha, so from the budget we can calculate that the yield response needs to be 4.1 per cent, or a yield total of 2.81t/ha, before the cost of putting on the extra nitrogen is covered.

Table 11.1 *Wheat gross margins*

Gross Income			***$/ha***
Yield	2.7 t/ha @	$140/t	378.00
Freight		$5.00/t	13.50
Total			364.50
Variable Costs			
Seed	89 kg/ha @ $0.15/kg		13.35
Fertiliser	70 kg/ha @ $420/t		29.40
Chemical Hoe Grass	1l/ha @ $22/l		22.00
Ally	5 gm/ha @ $1.34/gm		6.70
Machine Fuel			5.12
Maintenance			17.54
Insurance	$8.75/$1000		3.19
Total Variable Costs			97.30
Total Gross Margin			$267.20

Information from Departments of Agriculture indicate that given a reasonable spring, the required yield break-even is achievable. However there is a down-side as all the investment can be lost if the spring rain is poor.

What is the financial picture for investing nitrogen on 100 ha of wheat?

If nitrogen is applied, the farmer gets the highest expected wheat price and the yield response is 30 per cent, the initial $1500 nitrogen investment will return $5790. This is the best possible result. On the other hand if the yield fell 10 per cent, and the lowest expected wheat price occurred, the $1500 nitrogen investment would have contributed to an additional loss of $5145. This is the worst possible result.

Obviously while both the price of wheat and the cost of nitrogen are important, knowing what the growing season has in store and the likely yield responses will decide the success of the decision.

Our example farmer should therefore be putting effort into getting the best information available about weather forecasts and rainfall probabilities. If you feel long term forecasts are accurate enough, they also might be useful in decision making. Working out the probabilities for each seasonal event: rain, sunshine, prices, expected yield and so on, then weighing these up to arrive at some measure of probability might be a sensible move.

If price was thought to be very uncertain then farmers could consider fixing the price by using either forward contracts or futures contracts as mentioned earlier.

The risk of information comes when weather forecasts could be wrong or yield information may be inaccurate. The cost here for our wheat farmer could mean the loss of the initial nitrogen investment, which may then mean the difference between profit and loss for that year. Therefore, it is important to focus on the most sensitive information relevant to the decision. This will help cut the risk from making your decision.

12 External factors beyond the farmer's control

Some risks to the farm business are beyond the control of the farmer to avoid. The most important ones are exchange rates with foreign currencies, and government intervention with land and market rules.

Australian exchange rates, along with interest rates, were deregulated in the early 1980s. Since then we have seen wide variations in our exchange rates. The sensitivity analysis in Chapter 1 highlighted the importance of exchange rates on farm profitability.

Risk coming from the effects of law is another area of risk some people rarely see as important or significant. Laws reflect the attitudes and expectations of society. Generally new social attitudes are formed prior to new laws being made, so in most circumstances changes to laws can be foreseen. However, increased risk occurs when laws are changed with little warning.

Exchange rates

Exchange rates are unfortunately outside the control of farm managers. However understanding their impact on profits will help farmers improve their business management when exchange rates do fluctuate.

The Australian exchange rate is directly influenced by:

- ❑ the Federal Government's monetary policy (how the Reserve Bank of Australia sets interest rates)
- ❑ international market perceptions of how the Australian economy is performing
- ❑ speculators.

Federal Government

While the Federal Government is keen to see the market set the level of the $A against other currencies, it does not want too much fluctuation in the value of the dollar.

One way of steadying the $A is for the Reserve Bank to buy or sell the dollar to even out some of these fluctuations.

Another way to steady the $A is for the Federal Government to raise or lower interest rates. Higher interest rates attract overseas money into Australia, which increases demand for the $A. This helps strengthen its value. Lower interest rates usually have the opposite effect.

Setting interest rates (the Government's monetary policy) influences exchange rates but also affects other areas of the economy. This has caused problems between the Federal Government and farmers.

An illustration of this occurred in the late 1980s. The Government was concerned about an over-heated Australian economy. This meant general spending, including imports, was growing faster than the economy's ability to repay. To slow down spending and imports, the Government increased interest rates. This made it more expensive for people to borrow money which then encouraged spending to slow down.

This Federal Government policy tool is an indirect way to slow the economy and creates difficulties for farmers in two areas:

- ❑ the resulting higher interest rates attract more foreign funds into Australia and encourages our exchange rates to increase—making it more difficult for exporters to profit
- ❑ higher interest rates increase the cost of borrowing money

Fortunately for farmers, these trends go in cycles and policies set by Governments do change.

International markets

If our economy is seen by the international market to be strengthening, then foreign money also will come into Australia and strengthen the $A.

Signs of a stronger economy are lower inflation, low unemployment, an improving or a positive trade account and in Australia's case improvements in commodity prices. If these signs occur the $A will strengthen.

Speculators

The $A can fluctuate quite widely over a day. This swing is usually due to speculators trading the currency as they try to guess the shifts in exchange rates. This happens during Federal Government budget speeches, when inflation figures are released and when the Treasurer announces trade accounts.

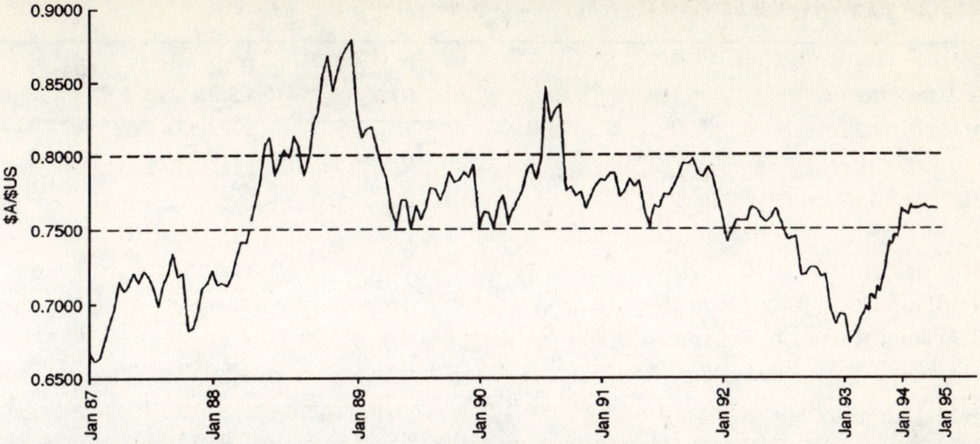

Figure 12.1 *Recent movements in the $A/$US dollar rates*

Shifts in the exchange rate

Figure 12.1 shows the recent shifts in $A/$US exchange rates. It is interesting to note the exchange rate has remained between $US0.75 and $US0.80 for most of the last five years.

The steady increase in the exchange rate from January 1988 to January 1989 was caused by the gradual increase in Australian interest rates, relative to the other major world currencies. The steady decline in the exchange rate since September 1991 was due to a combined result of falling Australian interest rates and the growing recession within the Australian economy.

The effects of exchange rate on farm businesses

As shown by the sensitivity analysis in Chapter 1, shifts in exchange rates can sometimes cause big swings in farm profits. However the most important exchange rate for farm profits is the one which occurs when the Wheat Board and Barley Board sell grain, when meat processors sell manufacturing beef to the US and Japan and, to a lesser extent, during wool sales. This happens because wheat, barley and meat are sold in $US. Most wool buyers come from Europe and Japan, and they are more interested in the price of wool in their own currencies.

Generally farm profits are greater when the $A exchange rate is lower.

In Table 12.1 wheat is used as an example. We might also have used raw sugar, manufacturing beef, or a host of other commodities. They are all affected in similar ways by swings in exchange rates.

Table 12.1 shows the effect on wheat gross margins when there is a 5 cent drop in the $A exchange rate compared to the $US. The fall in the $A from

Table 12.1 *Exchange rate effects on wheat gross margins*

		$US0.70 $/ha	$US0.65 $/ha	change
Gross Income				
Yield	2.7 t/ha @ $135/t	$364.50	$392.53	7.7%
Freight	$5.00/t	$13.50	$13.50	
	Total	$351.00	$379.03	8.0%
Variable Costs				
Seed	89 kg/ha @ $0.15/kg	$13.35	$13.35	
Fertiliser	70 kg/ha @ $420/t	$29.40	$31.66	7.7%
Chemical				
Hoe Grass	1 l/ha @ $22/l	$22.00	$22.84	3.8%
Ally	5 gm/ha @ $1.34/gm	$6.70	$6.96	3.9%
Mechanical				
Fuel		$5.12	$5.12	
Maintenance		$17.54	$17.54	
Insurance	$8.75/$1000	$3.19	$3.19	
	Total Variable Costs	$97.30	$100.66	3.4%
Total Gross Margin		$253.70	$278.37	9.7%

$US0.70 to $US0.65 would see wheat gross income increase by 8.0 per cent. However imports will become more expensive, which will affect high analysis fertiliser and to a lesser extent chemicals. The final results will see the gross margin improve from $254 to $278 per hectare, an increase of 9.7 per cent.

In this example, all high analysis fertilisers and half the chemicals are imported.

While this is a simplified example (as in reality chemicals and fertiliser would have been bought at a different exchange rate than when the wheat sales actually occurred), it does illustrate that an exchange rate has both a positive and negative influence on prices and costs. Generally a devaluation of the $A against the $US assists exporters.

Practical aspects of handling exchange rate risks

Unfortunately there is little that can be done to insulate farm incomes from the fluctuations of the $A exchange rate. Any exports are directly affected by exchange rate movements. Most domestic produce has to compete against imported products, where prices are also influenced by exchange rate movements.

However, some options are available to help insulate against such exchange rate movements:

- ❑ using the currencies futures market

- producing for a number of markets, of which only some are influenced by exchange rates
- keeping sound financial reserves.

Futures - a market tool

Industries such as wool, wheat, cotton, corn and sugar do have futures market options, where farmers can hedge their forward prices. Along with forward prices, the futures market also allows people to insure against moving exchange rates.

If farmers can decide the value of their production that depends on exchange rates and also know how much income is at risk if the exchange rate shifts, then it is possible to insure against these shifts. This is done by using the Currency Futures Market in Philadelphia USA. An option for $US50 000 (this equates to $A71 400 with a $US0.70 exchange rate) can be written and any movement can be insured. For more information, see a futures broker who understands rural business.

While these options are available, they are not commonly used by farmers because they are complex. Although useful in some instances, farmers need sound advice to use futures properly.

Diversifying

One option is to diversify the farm's production base into areas that are not subject to exchange rate movements. In some farming situations this is very difficult, as limited rainfall or soil types can restrict the range of agricultural options. It may be possible to become involved in domestic markets that are less influenced by exchange rate movements. Such areas may include:

- freshly-squeezed orange juice
- vegetables
- eggs
- fresh fruit
- dried fruit
- most non-farm income sources.

Take care when considering alternatives, as they will carry their own risks that you will need to take into account.

Keep a good level of equity in the business

Keeping a financially sound business by having plenty of equity or cash reserves is the most proven strategy against exchange rate fluctuations.

Keeping a high level of equity allows the business to withstand medium term financial shocks such as poor exchange rates.

Risk created by the law and governments

Government policy

There are a number of ways governments can influence farming risks and, with changes in government policy, new risks occur regularly. Recent government policies that have increased farm risks include:

- ❑ industry deregulation
- ❑ free trade
- ❑ using interest rates to manage the economy
- ❑ cutting government drought assistance.

Domestic deregulation of the wheat market has lead to variations in domestic prices.

Deregulating the finance industry has lead to greater fluctuations of interest rates. In turn, these fluctuations caused loan repayments to vary widely.

Using interest rates to slow the economy in the late 1980s also encouraged the $A exchange rate to increase, causing lower farm gate prices for exported commodities. Both Federal and State Governments except the Queensland and New South Wales Governments gave minimal financial support for drought affected farmers. All these combined factors have greatly increased farmer risks.

Land care

Over the last decade or so people have become increasingly aware of the importance of land conservation. The rise of the Landcare organisation is an example.

Farmers and the general public now see land as a finite resource that requires care and protection. Now land is getting more care and protection than it has had since white settlement in Australia. Today, society's perception is that our farmlands are degraded and must be restored and preserved.

Recently large amounts of government funding have been targeted for use in Landcare programs, which are assisting rural communities to improve their land.

Restricting the clearing of native vegetation is an example of recent laws designed to prevent further agricultural land development. These laws adversely affected some farms where land clearing was still in progress. While these farms are entitled to compensation due to restrictions on land clearing, the laws meant property plans were disrupted, creating added business risks.

The Mabo decision

The recent High Court decision on native land title, often called the Mabo decision, has created land tenure uncertainty amongst some Australian landowners. The issue of native title has now been resolved by an act of Federal Parliament.

The National Farmers' Federation has released information on what the Mabo decision means for Australian farmers and pastoralists. This information is summarised below.

Freehold land

One of the most important parts of the native title (Mabo) legislation is that freehold title extinguishes native title. That means farmers who own freehold land cannot have their properties subjected to land claims under the Mabo legislation.

Leasehold land

There are three classes of leasehold land; leases which may be invalid, valid leases where a statutory reservation of aboriginal interest already exists and valid leases where there is no statutory reservation of aboriginal interest.

- Leases which may be invalid: The Federal Government (Commonwealth) legislation will allow validation of any pastoral or agricultural lease which is found to be invalid, provided that lease was issued at any time prior to 31 December 1993.

  The act of validation will completely extinguish native title from the original date of the grant.
- Valid leases where a statutory reservation of aboriginal interest already exists. This is the case in WA, SA and the NT.

  Existing statutory reservations of aboriginal interest will be preserved and those existing aboriginal interests can not be extended except by State statute.

  The Federal Government (Commonwealth) legislation will leave open whether such interests remain statutory grants of the Crown, or are native title rights. This question is likely to be tested in the courts. Advice to the National Farmers' Federation, and the Federal Government is that existing aboriginal interests are not native title rights and exist only by statutory grant of the Crown.
- Valid leases where there is no statutory reservation of aboriginal interest. In the Mabo judgement, four of the seven High Court Judges expressed a general view that granting a lease extinguishes native title. Three judges expressed no view.

  Legal advice to the Commonwealth, the States, the National Farmers' Federation, aboriginal interests and the mining industry is that the valid grant of a pastoral lease is almost certain, at common law, to extinguish native title.

  It is possible that valid pastoral leases may be tested in court to decide whether any form of native title remains, even though there is no support for such action from mainstream aboriginal groups. Once a precedent is established, no similar actions will be heard by the courts. The Commonwealth's native title legislation establishes that the Registrar of native title claims would not accept vexatious claims, or claims over land where native title has been extinguished.

It is clear from the emergence of native title legislation that society's expectations about aboriginal rights are changing.

This uncertainty may affect investment in some areas, which increases the element of risk for a very few pastoralists.

Minimum cage size for egg production

There are some government regulations that set out the cage sizes for hens used for egg production. The industry has had these guidelines in place for some time now and cage sales have conformed to these requirements. However in recent times there has been debate as to whether these cage sizes should be increased. The main reason is due to concern by the animal liberation movement for the welfare of birds.

This pressure has increased the level of uncertainty in the industry, as changing cage sizes might require a large amount of capital expenditure. With the current low levels of profitability in the industry, this uncertainty has increased the risk to future profits. In some cases, these changes could even cause some businesses to become unviable.

Removing minimum grape prices

Until recently, the South Australian wine grape industry had minimum grape price legislation. This arrangement was set in place to prevent the private wine grape growers' prices from falling too low. The minimum grape price legislation was removed in the 1980s, creating a high degree of price uncertainty, as growers had to take the prices offered by the wineries. Removing the legislation increased those growers' risks.

It is interesting to note that in recent times Australia's wine exports have been booming and the outlook is for further growth. To create less risk, some parts of the wine industry have developed forward supply contracts to encourage increases in grape plantings. This is helping create a more stable environment. It is also giving grape growers more confidence in the industry and many are expanding their plantings. So there are emerging industry-developed strategies available to help minimise the extent of risk in the wine grape industry.

Dismantling the floor price for wool

The wool industry went through a boom period in the late 1980s, where both price and production greatly increased. This occurred under the control of a regulated market where a minimum price (floor price) was set for

Regulations on such things as cage size can be changed at any time

wool. Unfortunately the industry became greatly over-supplied. The reserve (floor) price was also kept above the price the wool trade was willing to pay. Wool was therefore bought by the Australian Wool Board and held in its stores around Australia. Soon a massive unsold surplus of wool built up. The surplus was so large, and was being added to at such a massive rate, that soon the whole reserve price scheme became unworkable. The inevitable happened and the Federal Government was forced to remove the floor price scheme.

Through these changes, wool industry risks increased.

Due to the floor price going, and the huge wool stockpile, the industry is currently experiencing very low wool prices, so the levels of financial risk for wool growers have greatly increased. These risk levels are so high that a number of growers are not surviving the industry changes.

The risks in the wool industry increased for two reasons:

- ❑ the interaction of market changes
- ❑ and Federal Government action in ending the reserve price scheme.

These ramifications and increased risks will be felt for some years.

Changes in taxes and tariffs

Governments have the habit of changing taxes—usually with little notice. There are many examples which affect farmers:

- ❑ changes to machinery investment allowances
- ❑ increases in wine and brandy taxes.

These changes in tax and duty systems cause quite a deal of financial uncertainty when they occur.

The change in how tax was collected on wine stocks greatly affected the cashflows of wineries. In this instance, the Federal Government did provide some warning of its intentions and phased in the new tax method, which meant the associated risks were greatly decreased.

Wine tax hikes in the 1993 Federal Budget were unexpected and badly hurt the domestic wine industry while benefiting the big exporters, as the tax increases only applied to wine sold in Australia.

Death duties

In the past, death duties were used extensively in Australia to raise government revenue. Even though they are currently not in use, farmers are always suspicious they may be reintroduced. The fact that there is this suspicion means there is an element of risk, which both rural businesses and accountants still consider in tax and estate planning.

Forms of ownership

Risk can be cut by altering the forms of ownership of a business.

An example is to form a company. The assets of the farm can be owned by the farmer and his/her family. Then the farm operations are run by the company which is a separate legal entity from the people who own the land and the house. The company owns neither the land or the house, but does own the machinery, and maybe the livestock. If anything goes wrong, and the company goes bankrupt, the land and house are free of debt and creditors can only recover assets from the company. They cannot touch the personal property of the company directors (the farm). Of course banks are less likely to lend money to a company with few assets but assets can be increased and companies can be built up gradually to cut risks.

Forming a family company to administer the property is one way to pass the farm on to the next generation without paying death duties

The Federal Government has been talking about reintroducing wealth taxes to take effect when assets are transferred from one generation to another. A simple way to avoid such taxes is to put the farm assets into a holding company, ideally a separate company from the one that runs the business. This way the directors may die, retire or otherwise change but the company continues on. Because the company endures there are no death duties or inheritance taxes payable on the company's assets. People can remain directors of family companies and still retain control until they die—the same result as passing a farm on to the next generation at death. This works because some farmers have four votes to the other director's one vote. That is the way they keep control of their company. Because company assets don't change ownership it is easy to therefore by-pass death duties and inheritance taxes. See your accountant and solicitor for advice, as this is one perfectly legal way of cutting risk from inheritance taxes.

Types of farming in certain catchment areas

Water quality is another important issue that will impact on some farms. The quality of water in the Murray River system is of great concern to all States that use Murray water. As a result, farming practices in the Murray catchment areas will increasingly be assessed for their impact on water quality. In some instances particular intensive agricultural practises may be prohibited.

Laws have already been passed to restrict the land use in parts of the Adelaide Hills to help maintain and improve the water catchment area for Adelaide's reservoirs. This has included restricting residential housing development, dairying and vegetable production. These restrictions have increased the financial risks to farmers in these catchment areas.

In Queensland the State Government has recognised the need for managing water in a different way. There, the State Government has put water resources under the Department of Primary Industries. They regard water as a primary resource.

In Queensland, water planning is run under an Integrated Catchment Management (ICM) program. Each river system is regarded as a separate land/water management area. That is the land, water, towns and the sea out from the river mouth are treated as a complete unit. Fertiliser put on a cane farm or a cattle property in North Queensland eventually finds its way to the Barrier Reef via the rivers that supply the towns with water. These nutrients affect farming, soils, water quality, river fishing and sea fishing and the tourist industry. This management system recognises river catchments as basic management units. There is risk to farmers who may be affected by ICM decisions, but since ICM boards make decisions about the catchments, and farmers are represented on those boards, the risks are minimal.

Practical aspects of handling legal and governmental risks

Managing risks in this area is very broad and is usually handled best off the farm. The measure of how well any management strategies are performing in this area are difficult to measure, as improvements can often be long term.

Politicians

The government makes the laws and politicians help frame the laws and policies for the country. One of the most effective means farmers have to change and influence governments is through political lobbying and keeping close contact with your local members of parliament.

This form of risk management tends to be more industry-oriented than farmer-oriented and progress often appears slow. However it is essential to be persistent to succeed when dealing with politicians and governments.

Joining a political party of your choice is another way of reaching and influencing politicians. It is surprising how politicians suddenly listen to farmers who are members of the party branches that decide who gets preselected to Parliament.

Agri-political groups

The task of lobbying politicians is generally beyond the ability of an individual farmer. That is why agri-political groups (like-minded groups of farmers) are necessary to influence governments. An example is the National Farmers' Federation.

Australian farmers have well organised agri-political groups whose prime aim is to lobby governments. By acting together more can be achieved than through individual approaches by farmers.

An effective strategy to cut governmental risk is to actively support your farmer agri-political organisation and actively take part in running it.

Community trends

One self-survival strategy is to be aware of community trends and be alert to any changes that may influence risks to your business. This will help guide your planning, and hopefully, minimise the effect of any likely changes that could occur in the future.

An example of this would be an egg producer who is planning to install new cages. The producer may be better off to buy the cages that will accommodate the new size specifications that may become law in any new cage size legislation.

Proactive planning

People can either react, proact or not react to a situation. People who react do so because a situation has occurred and they respond to it. People who proact do so because they see a situation will occur, so they prepare a response in advance.

In running a farm business we all react to the unexpected, but wise managers try to foresee events and proact to these so they are not caught unaware. Examples are being prepared for disease by vaccinating stock and maintaining machinery in the off season. This avoids problems of stock getting sick or having a harvester breaking down in harvest because no one checked and replaced worn drive belts in the winter.

Proaction is a strategy often talked about, but difficult to perform successfully. The assistance of other experts could help to position your business where it will be minimally influenced by any likely changes, for example selling a bush block before land clearing legislation becomes law.

Accountants are usually the best people to help with planning to minimise the impact of tax and they also try to forward plan to anticipate taxation changes.

Farm consultants can provide an idea of the profit and risk impact of any likely changes in the future.

13 Personal risks

Personal risks are those that directly affect a person's safety, health or well-being.

Historical background

The Industrial Revolution was characterised by increased mechanisation, a burgeoning growth of factories and the advent of machinery onto farms as workers left the land to take up jobs in towns and cities. This change first occurred in England then spread to the rest of Europe, the United States, Canada and Australasia.

Early machinery was fairly crude by today's standards and those early designs carried inherent risks in their use that would have horrified today's farmers.

Naturally accidents were common. People were caught up in tractor belts, pulleys, chain drives and so on. People were crushed beneath overturning tractors and run down by wheeled implements.

Many of these injuries were caused by the faster pace at which machinery ran compared to horses. Other accidents were caused by lack of training of farm workers. More injuries were caused by negligence and ignorance of what machines could do to the human body.

Accidents still occurred with livestock, falling from haystacks and similar events, but these risks had been a part of farming since time immemorial.

It was the advent of the machine on the farm, coupled with the rise of the first occupational health and safety legislation designed to protect workers in Australian factories that forced rural employers to improve farm safety standards.

Risk categories

There are several risk categories in which personal risks can be placed:

- ❑ physical risks—heat, radiation, noise and vibration
- ❑ chemical risks—carcinogens, dusts, poisonous gases, toxic chemicals and waste
- ❑ biological risks—insects, fungi, viruses and bacteria
- ❑ stress risks—work, physical, chemical and psychological pressures faced on or because of the job.

Causes and significance of rural accidents and occupational diseases

How healthy and safe are Australian farmers?

The city view of farming is that it is a healthy occupation and that farmers present a strong, macho image. However the partners of some 160 000 farmers and property managers know otherwise.

These partners remember the accidents, the injuries and the care they have given to their partners and employees and the physical support required when the injured person is laid up.

Farming is the third most dangerous industry after mining and transport, but we have not costed the economic loss across the whole of Australia.

Can you cost the economic loss to your rural business if you or your partner were killed or incapacitated? Could you afford to employ permanent or casual labour to replace you or your partner, even if you could find someone with the necessary skills and motivation to run your business? Are you insured to cover this loss?

Besides the drama of personal injury there is also a high incidence of occupational disease, such as deafness, dust related diseases and cancer. These conditions are often far more insidious than the more obvious deaths and injuries that accidents cause.

The incidence of skin cancer through long term exposure to the sun, added to the absorption of chemicals are among the other occupational diseases that affect farmers.

The social costs of work related injuries to employees and their families are often less obvious than the immediate economic impact of the injury, and these social costs are often borne by the community at large.

Therefore, occupational accidents and diseases present a major problem for rural industries, and the following figures provide an idea of the size of this problem.

Hearing loss

Of 381 farmers who had hearing screening tests at exhibitions and shows in several northern NSW centres in 1991, almost three quarters had mild, mod-

Many injuries are caused by the faster pace of modern machinery

erate or even severe noise-induced hearing loss (NIHL), while the rest had normal hearing loss for their age.

A similar farmer survey was conducted at the 1991 Yorke Peninsula Field Days in South Australia. This survey tested 128 farmers who were exposed to hazardous noise. The results showed that 84 per cent had hearing loss of some kind and 77 per cent had high frequency hearing loss.

Over 60 per cent of Australian farmers or graziers suffer from significant noise-induced hearing loss, from a wide range of causes. For example, exposure to the noise from an angle grinder, a chainsaw or a pig shed at feeding time for fifteen minutes or more. All these warrant the use of effective ear muffs or ear plugs.

Farm fatalities

There were 258 farm related fatalities in Australia during the period from 1982 to 1984. This represents a fatality incidence of almost 20 per 100 000 people per year in the farm workforce. This is more than twice the rate for all other occupations.

94 per cent of these deaths were male and 47 per cent were farmers or farm managers. Nearly 40 per cent of these deaths involved tractors or attached machinery.

30 per cent of these deaths were caused by other powered machinery, such as machine roll-over or being drawn into machinery like hay balers. About 70 per cent of the workplace deaths on Australian farms involve tractors or powered machinery.

97 per cent of all tractor related roll-over deaths involved tractors that were not, or probably were not, protected by a properly fitted and approved roll bar or cage, and/or the driver was not wearing a seat belt where provided. Many of these deaths occurred on reasonably flat ground.

On average one Australian farmer is killed by a tractor every four days.

A male agricultural worker has a 50 per cent higher probability of being killed in an industrial accident than an average worker. In addition to these deaths and injuries involving tractors and machinery, firearms cause an injury about every 23 hours throughout the year in Australia. Many of these firearm injuries occur on rural properties.

Hospital admissions

In South Australia from 1987 to 1988, there were 481 people admitted to hospital as in-patients from farm accidents. In 1989/90 this number had increased to 516.

16 per cent of the 1987/88 hospital admissions resulted from motor vehicle and other road vehicle accidents, 10 per cent were from accidents caused by agricultural and other machinery.

Nearly 14 per cent of these hospital admissions were the result of falls and ten per cent were caused by submersion in water, suffocations and injuries by foreign bodies.

The rural industry really is a most dangerous workplace—for example in South Australia it employs only 5 per cent of the total industrial workforce, yet is responsible for 17 per cent of all work related deaths.

About five employees in the Western Australian agricultural industry were injured every two days during 1987/88, so farming is also a dangerous occupation in Western Australia, as it is in all Australian states and territories.

Farming presents dangers for children

Children taking part in farm activities resulted in 13 of the 34 child deaths recorded on farms during a two year survey period from 1982 to 1984.

The other 21 children who died accidentally on farms were playing in a work area or were riding on farm machinery.

Several children under three years of age fell from tractors and were crushed under the rear wheel or fell into attached equipment such as a rotary hoe.

Mobile mechanical equipment accounted for 70 per cent of all deaths recorded in that survey.

Important rural risks

Ten major risk factors have been identified in farming and these factors apply equally to people who work and live on farms.

- ❑ Traumatic injuries—from tractors, equipment, animals, aircraft, slips and falls and flying objects.
- ❑ Ergonomic injuries—such as those caused by repetitive tasks, lifting heavy weights, and poor task, equipment or shed design.
- ❑ Biological risks—such as stings, bites, infections, infestations, organic dusts and infections from animal diseases (zoonoses).
- ❑ Physical risks—such as noise, vibration and electricity.
- ❑ Chemical risks—such as poisoning, cancer and burns from exposure to pesticides, herbicides, veterinary products and fertilisers.
- ❑ Climatic risks—such as heat and cold stress, exposure to solar radiation and problems associated with fire and flood.
- ❑ Psychological stresses—such as stress resulting from solitary work, economic hardship and hours and conditions of work.
- ❑ Attitudinal problems such as problems relating to risk taking, employer/employee attitudes, and cultural and language differences, especially for aboriginal and migrant workers.
- ❑ Lifestyle problems—such as alcoholism and the social isolation of farming.
- ❑ The lack of support services to rural communities—due to the wide

geographic spread of farmers, on the job mobility and itinerancy, restricting relevant information on rural occupational health and safety from reaching farmers.

Added to these risks are the facts that:

- most farms are family managed
- most farmers are self-reliant
- the risks that farmers face are considerably greater than those faced by most people in the workforce.

Perhaps the most insidious risk that faces rural people is stress, because stress builds up slowly and unnoticed over a long time.

Managing stress in yourself and the family

Stress is a universal human experience.

Unpleasant experiences can bring on stress. For example, losing stock in a drought, hail wiping out barley just before harvest, financial calamities and personal accidents.

These events are all potentially stressful to rural people.

Stress sometimes causes people to take dreadful actions against themselves and others so you need to be aware of the signs of stress in yourself, your family members and employees. By appreciating what stress is and by learning to recognise some of the early warning signs you may be able to stop someone's small problem growing into a big one.

What is stress?

Stress can be defined as a state that occurs when people are faced with demands from the environment that requires them to change in some way.

A simpler and perhaps cruder but more earthy definition in fashion at the moment is:

> Stress is the conflict created by civilised reason overriding a basic and primeval urge to beat the living daylights out of some sod who desperately needs it!

Whatever definition of stress you prefer, the clinical one or the more colourful one, stress is usually set up when some form of conflict arises in our lives—a conflict that forces change on us. Stress becomes a problem when we have difficulty coping with that change.

Stress is something that we all meet from time to time. Many people believe stress is the cause of most of their problems and must be avoided at all costs, but in fact, stress can have a positive effect and plays an important part in all aspects of our daily lives.

In the positive sense, stress is the reaction or reactions we have to events that can lead us to higher achievements. If we avoid this sort of stress we would not set goals, nor would we strive to achieve them.

In the negative sense, stress consists of all those forces that act on us to cause us discomfort or strain.

Whether we like it or not we have to learn to live with the effects of stress, because unless we can escape the wear and tear that daily living puts on our minds and bodies, then we cannot escape stress.

Here is an analogy of stress as it affects people.

Stress is commonly referred to in a physical sense, like engineering, as the strain that can be placed upon an object before it breaks. Each piece of metal or wood has its own breaking point. Think of an aircraft wing. If we subject it to the pressure it was designed for it will operate efficiently. Apply more pressure and the wing will flex or bend, which is an engineers way of saying the wing is showing stress. At this point the wing is still working, but not as efficiently. Put even more pressure on the wing and it bends even more. If we keep putting on ever more pressure the wing will eventually break. Then it is of no use to anybody—especially not the pilot and passengers.

We are all like that aircraft wing. If we learn to operate within our limits we will operate efficiently. However if we allow the loads placed on us to increase, we become subject to more and more strain. Then we stop operating efficiently and might even suffer a breakdown.

Just like metal and wood, we can all carry different amounts of pressure or stress. Stress therefore affects people in different ways. We will discuss these ways later in this section. However, it is important to note right from the beginning that each of us has our own level of ability to handle stressful situations.

For example, some farmers and business executives thrive on situations that would terrify others. Some people climb sheer cliffs or go parachuting. Others won't go near a cliff or touch a parachute.

Our reactions and our abilities to handle stress will always be different from other people's. Therefore you need to know where you stand with stress and how you should deal with it.

The fight or flee response

Stress has only recently been recognised as a contributing cause of many of the physiological and psychological problems we encounter in our lives.

The explanation for this is found in the fight or flee response.

Long ago our ancestors, when confronted by stressful situations, had two choices when faced by an angry mammoth or sabre-toothed tiger: either fight the animal threat or flee from it.

Both of these responses required some action. Our ancestor's bodies reacted to these situations by releasing hormones that were designed to give extra energy and increase their heart rates and blood pressure. These responses made sure enough blood got to the brain, the muscles and the other key organs to allow our ancestors to fight or run away. The method obviously worked because our ancestors survived long enough for them to beget a line of people that led to us. The hormonal reactions that occurred in our ancestors were designed to prepare them for action.

Today, those hormones act on our bodies in exactly the same way.

Today we have a problem. We still produce these body reactions, but because we are civilised we don't beat the bank manager over the head with a club when he or she has bad news about our overdraft, nor do we run away. We suppress our instinctive actions. Our actions are instinctive, because they are survival mechanisms that have been programmed into us over tens of thousands of years of evolution. (Those ancestors who didn't have a highly developed sense of fight or flight soon got eaten by sabretooth tigers and obviously they didn't pass on their genes to us.)

Doctors and psychologists believe that suppressing our natural and instinctive actions leads to many direct and indirect effects on our minds and bodies. One of these is the lowering of our immune response—and that means we are more likely to get sick.

To avoid these ill-effects we must:

- ❑ learn to identify the sources of our stress
- ❑ learn to recognise reactions to stress
- ❑ learn to find ways to deal with our stress.

The aim of avoiding these ill-effects is not to avoid the stress completely, but to learn to manage it and so minimise its harmful effects.

Sources of stress

Sources of stress are often called stressors. A stressor is a demand, situation or circumstance that disrupts our equilibrium and sets off our response to stress.

The following list contains a number of possible stressors. As you can see, some may annoy or irritate you immensely, while others may not.

- ❑ Uncertain economic conditions
- ❑ Uncertain weather conditions
- ❑ Poor or fluctuating commodity prices
- ❑ Lack of direction and communication for employees
- ❑ Unjust criticism
- ❑ Neighbour's activities
- ❑ Environmental problems
- ❑ Interference from bureaucrats
- ❑ Fear of hereditary diseases
- ❑ Fear of ageing
- ❑ Fear of rejection
- ❑ Noise
- ❑ Claustrophobia, aquaphobia etc.

- ❑ Fear of heights
- ❑ Uncertainty of the future
- ❑ People who have no appreciation of time
- ❑ People who lack consideration for others
- ❑ Indecision in others
- ❑ Bad drivers
- ❑ Spiders, insects, rodents etc.
- ❑ Family and relatives
- ❑ Change in lifestyle
- ❑ Competitive sports
- ❑ Gambling
- ❑ Dealing with incompetence
- ❑ Paying taxes.

It is important to note that one stressor in isolation may not cause any problem. The stress reaction sets in when a series of stressors are strung together.

For example you get up, its a poor day weather-wise and you wanted to do some outside work—say finish seeding a paddock as rain is forecast later. You get to the shed and the tractor won't start because the battery is flat. On the way to the paddock you notice a cow is having calving difficulties, so you stop and assist. Above all this activity is a nagging worry about whether this season will be any better than the last and if you can keep the bank overdraft from blowing out any further.

These events, or events like them, are all normal events in the daily lives of thousands of Australian farmers. When they are strung together over a long time they can take a toll on us all and possibly, before we even recognise it, we show signs of stress.

Signs of stress

We can show both physical and emotional reactions to stress. The following lists contain some of the signs that people show when they suffer from stress. These lists reveal the wide range of reactions people have and show how similar the symptoms are to other ailments. It is therefore a wise precaution to see your doctor if any of these symptoms persist.

Symptoms of mental and emotional stress:

- ❑ Temper tantrums
- ❑ Crying
- ❑ Laughing
- ❑ Lack of concentration
- ❑ Feeling inadequate
- ❑ Feeling irrational
- ❑ Being illogical
- ❑ Being forgetful

- ❑ Feeling depressed
- ❑ Being angry
- ❑ Being sad
- ❑ Being frustrated
- ❑ Relying on alcohol and other drugs.

Symptoms of physical stress:

- ❑ Impaired vision
- ❑ Back spasms
- ❑ Blackouts
- ❑ Increased heartbeat
- ❑ Increased perspiration
- ❑ Lack of appetite
- ❑ Increased appetite
- ❑ Insomnia
- ❑ Rashes—skin allergies
- ❑ Nail biting, fidgeting, twitching
- ❑ Butterflies in stomach
- ❑ Pacing the floor
- ❑ Screaming
- ❑ Ulcers
- ❑ Muscle tension
- ❑ Agitation
- ❑ Palpitations
- ❑ Asthma attacks
- ❑ Smoking
- ❑ Raised blood pressure.

Dealing with stress

The key to dealing with stress is to find the critical point of optimum stimulation where we are enthusiastic and active, yet not pushed beyond our own limits of performance.

We can suffer from the effects of stress from either having too much or too little to do. The trick is to find that area where we are operating within our limits and enjoying it. This level will be where we will operate at our best. Each of us must find this point ourselves, as it will vary from one person to another.

There are many ways of dealing with stress. Some of these are:

- ❑ relaxation
- ❑ recreation and exercise
- ❑ learning to cope with problems
- ❑ learning to cope with time.

Relaxation

When we are busy and under stress, relaxing is usually the last thing we think about, yet taking time to relax, even if it means something simple, having tea in front of the TV for an hour before going back out into the paddock. Relaxing is an important part of keeping on top of stress.

By practising relaxation we can break the stress cycle. Some ways we can relax are by:

- ❏ listening to music
- ❏ meditating

Relaxation and recreation are important in coping with stress

- ❑ muscle relaxation (deep breathing)
- ❑ controlling your time
- ❑ doing some things at a slower pace
- ❑ practising mental relaxation
- ❑ using visualisation techniques.

There are many ways we can become relaxed.

Attend a course or read a suitable book from a library and learn how to practise relaxation. Many techniques can be used whilst working to help break the stress cycle during the day.

Exercise and recreation

By keeping your body and mind in shape you will increase your ability to cope with the physical effects of stress. Taking regular exercise makes sure you keep yourself fit and your body in good condition.

Fortunately most people who work in primary production lead physically active lives so exercise is less important for farmers than it is for office workers. Nevertheless make sure you get the right type of exercise so you work all your body in a planned way. This can help relieve physical work related stressors that may themselves, contribute to stress. Talk to your local doctor when next you have a checkup—and get some advice.

Recreation gives you another interest, plus a form of enjoyment. Simply put, it takes your mind off other things.

Many people who work in primary industries take time off at weekends to relax, play sport, go out in the evenings and so on. However when things get difficult at home, for example if there is a drought on, or the crop doesn't look too good or wool prices fall, then people get worried and stop their relaxation activities. Typically we see people giving up their regular golf afternoon, or night badminton outing, as if somehow denying themselves that pleasure will help their farm situation. To be honest, giving up your badminton won't make it rain and certainly won't have any effect on wool, beef or banana prices. Giving up your badminton will certainly help raise your stress levels though.

It is especially important during times of stress that you keep relaxation high on your list of priorities. If things are bad, then probably your weekly round of golf, croquet or lawn bowls should be one of the last things you give up!

Both recreation and exercise have an important role to play in dealing with stress. Try to keep time in your weekly schedule for a little recreation and exercise and don't let stressors force relaxation from your life.

14 Management risks and management strategy

This book has discussed a number of risks that can influence farm businesses. The one major risk element is the risk of making management decisions. That is, some managers tend to make better decisions than others. These decisions then tend to minimise the influence of risk on the business.

It may appear that the progressive, innovative and entrepreneurial farms are the ones hurting financially through this current recession. This means that the decisions their managers made put their businesses in a higher risk situation. This may have caused their poor current financial position, not the nature of their enterprise. In these cases, it was management that created the higher risk for those farms.

Making correct decisions

Referring back to the farm risk framework in Chapter 2 note that the management component of the farm is placed in a pivotal position between external risks and business activities. It is the manager's responsibility to identify the important risks to the property, and choose how to best manage them.

The cost of an incorrect decision may be insignificant—such as which telecommunications system to use. However choosing which banker to use or whether to buy more property can have the potential to greatly increase business risk and hence influence the property's viability.

Risk will never be eliminated as unforeseen problems will always occur, but most leading farmers also claim they took risks early in their careers to lay the foundations for their current successful businesses. There is nearly always a balance between minimising risk and improving profit—and usually the more profitable options are the riskiest.

Deciding to make no decisions can also cause increased risks. For example there are opportunities within a cereal or export beef season where farmers and graziers can use the futures market to secure a forward price for their produce. Taking out an option to sell covers the price risk. Usually in each season there is a small window of opportunity to secure a favourably priced futures option. If farmers delay their decisions, then the opportunity to secure a sound price may be lost. In this instance, a decision not to make a decision or procrastinating will lead to greater business risks.

The SWOT analysis

A widely used technique used to scrutinise a business is called the SWOT analysis.

SWOT stands for:

- Strengths
- Weaknesses
- Opportunities
- Threats.

While this technique is used to analyse areas such as markets, competitors and business partners, it is also useful to perform the analysis on your own business and perhaps yourself. This will highlight your own abilities and show where you may need help.

Strengths

Strengths are those areas in which you know you have talent and excel. These strengths will have been noticed by others and you should already be using them. Examples of your personalstrengths may include:

- having a sound understanding of farming techniques
- being excellent with numbers and accounting
- being good at handling people
- being able to focus on goals
- being a good stock handler.

Weaknesses

Weaknesses are something some of us dwell on too much. It is important to understand your own limitations and have strategies that overcome these limitations. A classic example is the taxation system. Taxation law is continually changing, so we have accountants to assist us to minimise and complete our tax returns—hardly anyone tries to do this task without help.

Weaknesses in the farm business may include:

- ❑ lacking time management skills
- ❑ having poor computing knowledge
- ❑ having little understanding of farm chemicals
- ❑ having no long term business plans
- ❑ having difficulty communicating with bankers
- ❑ disliking extended fencing projects.

Opportunities

Opportunities here refer to those additional activities that are available to help increase the income base of your business. They can either involve the farm or be off farm activities. The opportunities for your business can be endless and it is a good exercise to sit down (with your business partner if you have one) and think of all the possible opportunities.

Examples of opportunities on the farm may include:

- ❑ improving cropping and pasture rotations
- ❑ introducing a new livestock type to the farm
- ❑ lowering debt by selling off a parcel of land
- ❑ developing an intensive agricultural enterprise
- ❑ using an improved cereal variety.

Examples of opportunities off farm could include:

- ❑ taking on some off farm work like shearing or teaching
- ❑ negatively gearing real estate
- ❑ setting up a share portfolio
- ❑ starting a buying co-operative
- ❑ investing in a local business.

Some very ingenious ideas have worked for farming families to increase income and you or your neighbours may have already done this successfully.

Threats

Business threats are those events that may devalue your business. This book on risk has touched on a number of these threats. You may also have some local threats that could influence your property. For example, the local council could be considering restricting further development of intensive agriculture, such as piggeries and poultry farms. If you are a poultry producer this may mean you may not be permitted to expand in the future. It could also mean you will have to relocate your business.

Other threats may include:

- ❑ land clearing restrictions
- ❑ the emergence of herbicide-resistant weeds
- ❑ the finance company's repossession of a major piece of plant—like a tractor

- ❑ the threat of interest rate increases
- ❑ bad weather
- ❑ drought.

Examination

Look through your SWOT analysis answers and study them carefully. The exercise of performing the SWOT analysis on your business or yourself can be quite threatening. However, there are also important benefits:

- ❑ recognising your strengths
- ❑ capitalising on your strengths
- ❑ knowing when to call on additional help to assist with vital activities and decisions.

The SWOT analysis also can help strengthen your property management practices and help cut business risks.

Your goals and reasons for farming

Another powerful resource you can use to help manage risk is goal setting. There are a number of valuable books and resource materials available on the topic.

Setting goals ensures you keep your activities on track and helps focus your efforts on achieving your goals. It is important to define what business aims you are seeking. This could also include family, personal and community goals. These goals could include the following:

- ❑ have a business with a return-to-capital-invested of ten per cent after tax
- ❑ buy a new tractor before next season
- ❑ reduce debt by $20 000 a year
- ❑ successfully establish your children on the property
- ❑ have four weeks holiday per year
- ❑ improve farm profits by 10 per cent this coming year.

The best way to begin goal setting is to brain-storm and write down all the things you wish to be, to do or to have. Let both your mind and pen run freely. Fill up many blank pages if you wish. With brain-storming it is important not to let reality get in the way of ideas. Brain-storming works by taking all ideas, both realistic and fanciful, and sifting through them for possible schemes. Eliminating way-out ideas too early can often close off unintended options. It would also be beneficial to do this brain-storming exercise with your spouse if appropriate and see whether your aspirations differ.

It is important these goals be both personal and believable to you—even if some might seem far-fetched. Now divide these ideas into short, medium and long term goals.

Short term goals are those goals you wish to achieve within the next twelve months. They could include refencing the bottom paddock, laying new water pipes from the bore to the first water point or replacing the ute.

Medium term goals are those you wish to achieve in the next two to five years and could include larger goals like reducing debt, replacing machinery and putting the kids through boarding school.

Long term goals are those you wish to achieve in five to ten years or longer. They might include both large and small goals, such as successfully shifting the farm's management to the next generation, establishing the older generation off the farm for retirement, taking that overseas holiday you have always been talking about and buying more land to enhance the farm's viability.

The most important aspect of goal setting is to write down why you want or need to achieve these goals. For each goal outlined above, write down five reasons why you want or need to achieve that goal. Having strong reasons will give you an optimistic outlook on achieving these goals. When things go wrong with your achievement schedule, an optimistic outlook can help keep you going.

Is management your biggest business risk?

This question is one of the most difficult to answer as most managers do not find it easy to look inward and ask this question of themselves. The answer is very personal and you may have to find a friend and mentor to help you decide on an objective answer. Unfortunately the real answer will only become evident, as you go about trying to achieve your goals.

Completing the SWOT analysis and your goal planning exercise will be rather time consuming to do thoroughly. Nevertheless, it will give you a good idea of where any management weaknesses may lie.

Developing a risk strategy

Having appraised both your own management and your property's position, you can now develop an overall risk managment strategy for your operation

Look at the specific risks covered in this book:

- ❑ Weather related risks
- ❑ Drought risks
- ❑ Risks from short term commodity price fluctuations
- ❑ Risks from long term commodity price fluctuations
- ❑ Asset risks
- ❑ Financial risks
- ❑ Biological risks

- ❑ Land degradation risks
- ❑ Risks from inadequate information
- ❑ Exchange rate risks
- ❑ Legal and governmental risks
- ❑ Personal risks
- ❑ Management risks.

Make a list of all the risks that could affect your business under the headings above. Next, rank them in order of importance, putting the most important risks first.

You should now have a list that covers the greatest risks to your business.

It will not be possible to address all your risks immediately, therefore focus your efforts on those areas where you can have the greatest impact early on. Select the five greatest risks. Then go through the following procedure for each.

- ❑ Describe the risk: this will help you appreciate the nature of the risk.
- ❑ How could this risk physically affect your business? This may not be applicable for all risks but beware not to overlook possibilities such as your tractor being repossessed.
- ❑ How could this risk financially affect your business? If the worst event happened, how much would it cost you?
- ❑ List all the strategies you have available to help minimise this risk.
- ❑ Which management strategies will you use for your business?
- ❑ Under what conditions, or trigger, will this strategy be implemented?
- ❑ How are you going to activate this risk management strategy? Put a realistic and achievable time limit on implementing the different elements of this strategy.
- ❑ What resources will you need to implement this strategy? These could be fencing materials for dividing a paddock or advice on futures markets or share brokerage.

Long term plan

The above exercise should help you focus on what you need to do to minimise each risk. Now place all your risks in perspective and plan your risk management strategy for the coming year.

Write down your risk management activities for the next twelve months.

It is important to place these activities in a realistic and achievable time frame.

References

Daniels Peter, 1985, *How to reach your life goals*, Gillingham Printers Pty. Ltd.

Drought Policy Review Task Force, May 1990, *National drought policy*, Australian Government Publishing Service, Canberra.

Rohn Jim, 1989, *Seven strategies to wealth, health and happiness*, Prima Publishers.

Index